# Greening the Ivory Tower

## Improving the Environmental Track Record of Universities, Colleges, and Other Institutions

Sarah Hammond Creighton

The MIT Press
Cambridge, Massachusetts
London, England

This book was set in Sabon by Achorn Graphic Services, Inc.

Printed and bound in the United States of America.
Printed on recycled paper

Library of Congress Cataloging-in-Publication Data

Creighton, Sarah Hammond.
Greening the ivory tower : improving the environmental track record of universities, colleges, and other institutions / Sarah Hammond Creighton.
p. cm. — (Urban and industrial environments)
Includes bibliographical references and index.
ISBN 0-262-53151-8 (pbk. : alk. paper)
1. Environmental management—United States. 2. Universities and colleges—Environmental aspects—United States. 3. Green movement—United States. I. Title. II. Series.
GE310.C74 1998
363.7'0576—dc21 97-39382
CIP

# Contents

# Tables

# Figures

# Boxes

# Foreword

In 1988, ten environmental-planning graduate students at the University of California at Los Angeles began meeting to decide on a common topic for a group comprehensive project. The project, a client-based research undertaking, served as a substitute for the students' master's theses and was considered the signature event in their professional training–oriented education. Two topics were considered: an environmental justice theme identifying and evaluating the complex of environmental hazards experienced by residents of East Los Angeles, and an evaluation of the environmental issues associated with the activities of UCLA. The students essentially divided between those who considered community-based justice or "risk discrimination" issues to be of paramount importance and those students who felt a community imperative to take responsibility for what they called their own backyard. The ten students eventually split into two separate groups, both producing high-quality projects.

The award-winning UCLA-related study, In Our Backyard: Environmental Issues at UCLA, Proposals for Change, and the Institution's Potential as a Model, was notable for several reasons. First, it demonstrated that the university, partly as a function of its size, did indeed generate significant environmental impacts: water use (eighth largest user in the city of Los Angeles), energy use (third largest user in the city) air emissions (tenth largest emitter of carbon monoxide in the regional air basin), sewage flow (accounting for nearly one-half of 1 percent of the wastes treated at the City's Hyperion Treatment Plant), hazadous materials use and waste generated (more than 100 tons annually), solid waste sent to landfills or incinerators (15,000 tons annually), and so forth. Second, despite some interest in environmentally oriented programs (a ride-share program, use of alternative fuels for some campus vehicles), there was an absence of environmental efforts in several key areas as well (continuing emphasis on the automobile by investment in parking infrastructure, absence of recycling or reduction strategies, little support for alternative transportation strategies such as bicycles). Most symbolic of this absence of effort had been the failure to establish a recycling program. During the 1970s, after the initial enthusiasm of the first Earth Day, a modest student-led recycling program was begun. This was soon suspended due to the chancellor's decision to remove the recycling bins since he felt that the bins spoiled the aesthetics of the campus. Ultimately, the response of the university administration to the In Our Backyard study was also negative (primarily due to the study's extensive press coverage), and the opportunity to follow through on this evaluation, or environmental audit, was never pursued.

Despite this reluctance at UCLA, the In Our Backyard study in fact became one of the keystone documents that helped inspire what has since become known as the campus environmental audit. These audit activities—some undertaken in conjunction with university administation support, some forced to operate independently or at the margins (often as student activist–oriented investigations)—have quickly extended to more than several hundred campuses in just a few years. But despite widespread interest in the concept, the campus environmental audit has remained largely ad hoc, often disjointed, and has often been lacking in breadth and analytic rigor.

Sarah Hammond Creighton's work in this area is an important and welcome addition to this literature. Derived in part from one of the most ambitious university environmental programs, based at Tufts University, Creighton's study captures the range of issues necessary for any comprehensive evaluation and provides a framework for investigation and analysis that can turn such an effort into a more comprehensive and coherent undertaking.

For the campus environmental audit to identify the issues and help transform a university environment, it needs to become both subject and object of rigorous environmental analysis, emerging in effect as a program for environmental literacy as well as environmental improvement.

The campus environment needs to be seen as an area where the relationships between work and knowledge constitute the basis for environmental change. And the campus environmental audit as an instrument for such change must also firmly establish the principle that environmental literacy—and environmental justice—is a function of responsibility for those places where we work as well as live and play.

Robert Gottlieb

# Acknowledgments

There are many people who helped to make this book possible and contributed some of the ideas or helped me to learn the lessons that I have tried to share here. Friends and colleagues at Tufts University were important to the successes there and to the increased knowledge of greening a campus. Tony Cortese had the vision to seek support for campus stewardship efforts. Bill Moomaw led the Tufts CLEAN! project in its infancy and still continues to push the envelope of what seems realistic and possible at Tufts. Members of the Tufts community, in dining services, purchasing, physical plant, and elsewhere, took (and continue to take) suggestions and ideas, trying them out and thereby providing me with rich experiences that form the basis of this book. Graduate student researchers and colleagues helped me to gather information for Tufts CLEAN! that I later used for some of the background for this book: Peter Allison, Karen White, Caroline Ganley, Chris Bell, Meg McClennan, Priscilla Ryder, Lucy Edmondson, Jim Greenbaum, Ilze Gottelli, Sheila Machado, Warren Goldstein-Gelb, and Maureen Hart. Special thanks to Patti Lee, Molly Anderson, Eric Friedman, Keith Kidd, and Ann Rappaport for their thoughtful comments on early drafts of chapters and for their dedicated work to green Tufts. Special thanks to H. Emerson Blake for his wisdom, guidance, and editing. Many thanks are also due to my husband and family who put up with this project longer than we wanted, provided moral support through the project, and pushed me to see it to completion.

# 1 Introduction

In 1990 the U.S. Environmental Protection Agency awarded a research group at Tufts University's environmental center a grant to undertake an effort, known as Tufts CLEAN! (Cooperation, Learning, and Environmental Awareness Now!), to reduce or eliminate the environmental impacts from the university's own operations. Tufts was chosen because of its leadership role in environmental education and research and its commitment to environmental programs. The small group of researchers and students who gathered to work on the project recognized that universities use resources and generate a great deal of waste in conducting their business and therefore offer a multitude of opportunities to prevent pollution, demonstrate clean technologies, and take leadership for environmental protection. Colleges and universities are microcosms of society's systems to house and feed people, conduct research, and administer programs, so their operations have many of the same consequences and opportunities for the environment as homes, offices, restaurants, and hotels.

The Tufts CLEAN! project could have underwritten the installation of a technology that would reduce the university's environmental impacts in a single area, such as an efficient new heating system or a fleet of electric vehicles that would reduce on-campus air emissions. Instead the team chose a more comprehensive approach by serving as a resource and catalyst for action to reduce the environmental impacts of many of Tufts' own activities on and off campus. The team examined specific issues in depth, such as food waste, transportation, energy efficiency, and procurement practices, and members worked with target departments to develop recommendations. Tufts CLEAN! differed from many other university environmental action efforts in that team members worked directly with the university staff in operations departments such as buildings and grounds (known on some campuses as physical plant or facilities and maintenance), purchasing, dining services, printing services, and computer services, as a catalyst for action and an information resource to include environmental factors in the decision making of department personnel.

The team members—academics, staff researchers, and students—had little background in running a project of this magnitude and initially relied on a descriptive approach to quantify the needs and identify solutions based on empirical evidence. Later, team members spent a great deal of time understanding practical considerations that are essential to keeping the university running smoothly. The team engaged students to help on research related to purchasing, dining, energy, composting, and hazardous materials. In all but the first year, I served as project manager, and the team included several faculty members who participated on a limited basis.

Throughout the first year of Tufts CLEAN! the project team gathered data and made recommendations about the logical and justifiable changes in practices and policy that the university could make in order to reduce or eliminate its burden on the environment. In many cases, the recommendations have been implemented; however, it took a long time to recognize that as a research team, we were nearly powerless to effect the changes in university systems and procedures necessary to implement the far-reaching waste reduction and conservation goals we envisioned—for example, solar hot water systems, natural gas–powered vehicles, recycled paper with 100 percent postconsumer waste, and reusable dishes in all dining halls and cafeterias. For example, members of a research team do not upgrade the lighting; that is done by the buildings and grounds department. Decisions about the cleaning products used at Tufts are made by the head of custodial services and the purchasing department. The grounds department waters the athletic fields, and dining services ultimately decides whether the students eat off paper, polystyrene foam, or china dishes.

Perhaps the biggest lesson from Tufts CLEAN! is that actions to reduce or eliminate a university's adverse impacts on the air, land, water, health, and safety require the personal commitment and direct involvement of university staff who have the responsibility for operating the institution on a daily basis. This lesson is essential to the success of broad-based university environmental stewardship on any campus and is one important reason for writing this book.

## About This Book

This book is about college and university actions to reduce environmental impacts from campus decisions and activities. It is about the numerous small and humble actions that the members of a university community can undertake to reduce the environmental footprint—the multitude of on- and off-site impacts resulting from university business and decisions—of their institution. The book is also about the process of planning and undertaking these actions at a college or university, in the hope that the process will lead to successful outcomes. Although the Tufts experience forms the basis for much of the thinking behind this book, the actions described go well beyond those that Tufts has undertaken, and, examples from other institutions are used to illustrate successful implementation actions.

This book is written for people who, like me and members of the original Tufts CLEAN! team, are interested and motivated to help green their campuses but have little or no experience with changing institutions or with the technologies that are needed to accomplish the task. Students, faculty, staff, and administrators will find strategies that are relevant to them, as well as ways to support and encourage action throughout the university.

The environmental stewardship actions that are detailed in this book are important steps toward transforming a university to an institution that treads lightly on the earth. The book offers numerous action steps for greening the ivory tower, and each action is simple enough that any university community can expect to be able to accomplish it. The truly green university will need to undertake comprehensive implementation of these actions, and go beyond those I describe. If only a few actions are taken at each of the more than three thousand institutions of higher learning in the United States alone, progress will be made. Undertaking all the actions will be overwhelming if each action is considered individually;

instead, these actions are concrete examples of how to operationalize the concepts of reducing waste, decreasing hazards, maximizing efficiency, handling wastes optimally, and engaging members of the university community in the process. Universities that are committed to a goal of environmental stewardship will find that university commitment supports and reinforces the individual actions that become expectations rather than exceptions. On the other hand, members of many institutions will find that they need to approach these actions individually, because overall commitment or a master plan for environmental action is lacking.

**University Environmental Impacts**

Colleges' and universities' use of electricity, oil, natural gas, water, and chemicals is significant and may be the largest uses in the community or region where an institution is located. Tufts' main campus, for example, uses more electricity than any other business in its electric company's district. The impacts of energy use by colleges and universities are probably these institutions' largest single environmental impact. A small town may find that its college is a major contributor to the wastewater treatment plant, as well as the greatest single user of clean freshwater. Even if the university operates its own wastewater treatment systems and supplies its own drinking water, the financial costs of running these systems, complying with laws for wastewater treatment, and providing safe drinking water are large. Universities and colleges generate large quantities of trash, or solid waste, as well as hazardous wastes, such as chemical waste, pesticides, paints, solvents, and radioactive wastes. In Massachusetts, university medical schools and research labs are among the largest generators of low-level radioactive wastes, stemming largely from the research laboratory. Hazardous chemicals used in laboratories, pesticides, insecticides, and fertilizers are abundant on college campuses and contribute to water pollution and indoor air pollution problems that can endanger workers, students, and community safety and natural systems. In New England alone, thirty-five universities or colleges are listed as contributors to listed Superfund (hazardous waste) sites for their failure or the failure of their contractors to dispose of hazardous waste properly.[1] Chemicals that deplete the ozone layer, causing increased human, animal, and plant exposure to harmful ultraviolet radiation, are prevalent in cooling and refrigeration systems, fire extinguishers, automobiles, and libraries. Disposing of the university's solid waste also contributes to air and water pollution. Transportation to and from the campus by deliveries and commuters may increase noise, congestion, and air quality problems for local communities.

Significant indirect impacts are created off campus by the use of services or the production of goods outside the institution, such as pesticides on food served on campus or dioxin resulting from bleaching paper used in offices. These environmental problems may seem distant, but they are indeed related to university actions. The purchasing and product use decisions that members of the university community make individually and collectively may influence off-site growing and manufacturing practices, the demand for environmentally friendly products, or the reduction of waste and pollution.

The following statistics provide a glimpse into Tufts' annual activities and environmental impacts on its main campus:

- 3,200 students housed by the university
- 5 million meals served
- 14 million copies made
- 65 tons of paper towels purchased
- A large-quantity generator of hazardous waste[2]
- 110 million gallons of water used
- 2,127 parking permits issued
- 1.1 million gallons of fuel oil burned in four central heating plants, resulting in the emissions of 22 million pounds of carbon dioxide
- 23 million kilowatt-hours of electricity consumed, resulting in the emission of another 34 million pounds of carbon dioxide
- Over 2,000 tons of solid waste generated

## Motivation for Campus Environmental Action

Institutions of higher learning teach young people the professional and intellectual skills they need to cure disease, run businesses, lend money, and legislate policy. Universities also conduct basic research that is instrumental in understanding the natural world and our effect on it. In teaching these skills and investigating new research areas, universities and colleges have a unique opportunity, in the classroom and by the example of their physical plant, to provide students with a basic understanding of the interaction between business decisions and the natural systems on which our health and well-being depend. Furthermore, universities have an opportunity to make choices themselves and become stewards of environmental systems in anticipation of the needs of future generations. Universities can both teach and demonstrate environmental principles and stewardship by taking action to understand and reduce the environmental impacts that result from their own activities.

Many of the actions that reduce the environmental footprint of the university involve the reduction of waste: wasted heat, wasted water, wasted electricity, wasted chemicals, and solid waste. Thus, actions to reduce these wastes represent opportunities to save. Embracing waste reduction projects can save a university money, although implementing some can require up-front capital. But the fact that most institutions of higher learning plan to exist well into the next century makes long-term thinking and investment in long-term waste reduction prudent. Furthermore, the long-term costs of solid and hazardous waste disposal, freshwater, wastewater treatment, and energy will continue to rise, making efficient use of these resources a financially sound decision. Liability from poorly handled wastes or accidents to students, employees, and property can also be costly, and they are avoidable.

Since universities are generally long-lived institutions, they should be concerned with the long-term health and livability of their community and region. A beautifully maintained campus surrounded by traffic, air pollution, litter, and a filthy river will have difficulty attracting students. Furthermore, environmental efforts can be a selling point for the university, both within its community and with prospective students. Nearly 20 percent of the students who enter Tufts list environmental concerns as one of their three top concerns.[3]

## University Environmental Action

Some universities and colleges are already implementing prudent practices that comply with or exceed local, state, and federal regulations, improving energy efficiency, recycling wastes, and improving the storage and handling of hazardous chemicals. Student audits of their campuses have raised awareness of their university's contribution to local pollution problems and have spurred action. Across the country, students have started recycling programs, advocated for divestment from corporations with poor environmental records, and helped administrators and staff to research and implement programs to buy recycled products or begin composting. Faculty members have expanded curriculum to include the study of resources and pollution, as well as the policies, and engineering that contribute to both problems and solutions. Facility managers have discovered the rapid payback of energy-efficient lighting. Dining service staffs routinely separate cans, bottles, and cardboard for collection and recycling. These efforts and the hundreds of others like them represent progress toward a greening of the ivory towers that teaches our future leaders and workforce.

There are many exciting and innovative efforts in nearly all sectors of campus operations and activities.[4] Sometimes a particular effort will flourish in one part of the university, but the other operational units of the same institution will lag far behind. In many places, efforts have started but dwindled; in others, seemingly logical actions remain undone. Few, if any, colleges or universities have undertaken comprehensive, across-the-board environmental stewardship in curriculum, operations, and university policies. A number of institutions, including Brown University, Ball State University, Georgetown University, the University of Vermont, the University of Wisconsin, and Tufts, have taken steps to think comprehensively about the range of stewardship actions and to embrace them throughout the institution. As at other campuses, Tufts' successes are found in targeted areas where individuals are willing to spend the time; seek out the information and support of government agencies, nonprofits, knowledgeable alumni, and other helpful resources; or change the way they do business. Sometimes these efforts are motivated by outside pressure, laws, mandates, or incentives. Other efforts are motivated by student pressure or financial opportunity. Still others stem from personal or institutional commitment to making the world a better place.

Some universities have yet to begin, but on many campuses the easiest initiatives are already underway—those with opportunities for quick financial payback, positive visibility, research funding, or student participation. In the next phase of the campus environmental stewardship movement, colleges and universities will see the long-term benefits and invest in capital projects with longer-term paybacks, change curriculum to reflect holistic thinking, reduce or eliminate hazardous by-products, and examine each business decision in the light of the quality of life and the quality of the natural world.

As with all other institutions, universities are communities of people: students, faculty, administrators, and staff. To date, no university or college community has completely understood and acted on the opportunities to save money, reduce risk, demonstrate new technologies, and increase student learning that a truly green university might offer. The progress toward that end will require the commitment of many individual people. It will involve rethinking some priorities, taking risks, making mistakes, and persevering.

## The Green University of Tomorrow

The green university of the future may use resources efficiently, create little or no waste, and take full responsibility for any waste that it does generate. As Bates College has begun to do, the green university may purchase organic foods grown by local farmers. This university will invest its endowment to purchase shares in companies that specialize in efficient technologies rather than polluters that destroy precious lands and waters to provide stockholders with ample returns. As Rutgers University has begun, the green university will rewrite contracts to favor reuse and recycling and to buy products from green businesses that have taken steps to reduce their own environmental impacts. The green university might renovate an existing building rather than build a new one, or reduce mowing and increase wild plant species in target areas, as the University of Buffalo is trying. The University of Waterloo allows students in some classes to submit papers electronically, and perhaps these computers will one day be powered by local and renewable sources such as solar or wind. Further, this green university will include learning and appreciation for the physical environment and our connection to it in courses, laboratories, and university culture and throughout the institution's physical plant.

The complexity of our lives and institutions makes this vision nearly impossible unless change is seen in some of our goals, expectations, rewards, educational processes, and perhaps even the tenure and promotion system—long-term and ambitious goals, to be sure. In the near term, universities can make progress toward this vision. Members of university communities can learn how to reduce or even reverse these impacts through their individual and collective actions. The green university recognizes that it has a responsibility to lead rather than follow, try new solutions to old problems, and continuously improve its business of environmental protection.

## In Summary

Tufts CLEAN! made progress on environmental stewardship projects throughout the university. Perhaps most important, Tufts continues to embrace many initiatives that Tufts CLEAN! helped to begin, such as hiring an energy manager, upgrading the lighting with efficient technologies, regularly purchasing energy-efficient computers, recycling and composting wastes, exploring the use of organic foods, and displaying a dramatically increased awareness about environmental issues and the university's role in addressing them.

The Tufts approach encompassed master planning and a vision but emphasized implementation. Tufts CLEAN! taught several major lessons that are essential to success of environmental initiatives in a decentralized place like a university and are very transferable to other settings. We learned that the business of *greening* a place requires sound research, attention to details, and unfailing commitment. We learned that university operation staff in purchasing, facilities, dining, printing, computer services and other departments hold the key to implementing many projects. We learned that to maximize student energy and faculty expertise requires careful targeting of those resources. We learned that decisions that affect the environment are complex and that the environmental efforts must complement rather than consume the educational mission of the university and its departments.

# 2

# Making Environmental Change Happen: The Essential Ingredients

Regardless of whether a university is on the forefront of environmental stewardship or just beginning to carry out a few isolated changes to protect the environment, there are many opportunities to improve the university policies that affect the environment. Often the actions needed require that people change routines they have used for years, and this makes these actions inherently difficult. Changing habits requires information, commitment, encouragement, and feedback. And these key factors require the time, energy, and enthusiasm of individuals if environmental stewardship is to spread throughout the university. The thinking behind decisions must be flexible to make significant progress in reducing the environmental footprint of any institution.

Five ingredients are key to successful university environmental action: (1) understanding how the institution works, its players, and its decision making; (2) university commitment and demonstrated support for environmental action, often articulated in an environmental policy; (3) a university-wide environmental planning committee or smaller issue-specific committees; (4) individual leaders; and (5) an understanding of the basic principles of environmental protection. Not all university efforts have all of these components, but the most successful efforts, especially those making progress on more than a single isolated project, include many of these institution-wide support systems.

The business of greening the university also requires careful and directed gathering of data that contribute meaningfully to change. Finally, greening the university requires attention to the economic realities of proposed actions, an acknowledgment of existing university priorities, a willingness to try projects on a small pilot scale, attention to publicity, and an understanding that priorities may need to be set along the way.

This chapter provides the leaders of a campus environmental effort with background information about the way universities work and some of the elements common to successful campus environmental stewardship projects.

## How the University Works

To effect change—for example, to increase the use of recycled paper or improve the pest management practices on campus—campus environmental leaders must understand the way that universities make decisions. Some colleges and universities have a strong sense of community, others are bureaucratic, and still others make decisions based on internal political power.[1] The better that campus environmental leaders understand how the institution is organized and makes decisions, the more likely that they will be effective.

Universities and colleges that have existed as institutions for many years have well-established systems of operating, and they tend to be fiscally and operationally conservative. Moreover, they are the purveyors of ideas and the teachers of classic methods, history, and a body of knowledge that is often more theoretical than practical, making these institutions less than adept at the pragmatic thinking needed for implementing campus environmental action. Although by virtue of their role as research institutions, some thinking at colleges and universities is innovative and experimental, but little of this seems to rub off on the institutions themselves, which are often run much the way they have been for years.

### University Players and Decision Making

Five major groups make up the university community: the trustees and overseers, the administration, the faculty, the staff, and the students. The surrounding community may be an important sixth player but typically is considered an external influence rather than a member of the university family. Environmental stewardship at the university requires understanding the roles and interests of each of these groups. For example, a common mistake on many campuses is to fail to develop action plans in concert with university staff. Students sometimes demand action rather than finding ways to use their time and energy to research or pilot-test options. Faculty may see the implementation of environmental concepts as inappropriate work and become frustrated when change is slow or nonexistent. Working together for environmental stewardship requires developing an understanding of the needs and expectations of each member of the university. For example, students must take final exams, and grounds managers are rewarded for a campus that looks like a country club. A cooperative atmosphere is essential for successful change.

**Trustees** The board of trustees is responsible for selecting and managing the president and for determining the institution's mission, long-range plans, and educational program.[2] It is responsible as well for the university's financial health, community welfare, fund raising, and arbitration of disputes. Although historically boards were intimately involved with running the university, they are increasingly less involved,[3] so it is rare that boards of trustees are actively involved in greening the university, although their involvement will depend on the character and experience of a particular university's board.

The long-term nature of the work to protect resources, save money, and build on the innovative educational opportunities that the environmental field offers should put these issues high on the agenda of a creative board. Trustees may be interested in the competitive and financial advantages of environmental programs, and their attention to these programs can bring prestige and credibility to the efforts.

**Administration** College and university administrators—the president, vice presidents, provost, and the deans—are responsible for the multitude of day-to-day demands that come with running an institution. In addition to fund raising and alumni relations, they are attentive to community affairs, public relations, housing, student discipline, and the concerns of faculty. Administrators are particularly interested in how environmental initiatives can complement other university initiatives, such as those with financial or public relations benefits. Universities that have a high-level administrator who takes an interest in the environmental action on campus tend to have greater campus-wide participation and cooperation than schools where the administration is ambivalent or reluctant.

**Faculty** The faculty is responsible for teaching, student advising, and research. Many faculty have job security offered by tenure or are in tenure track positions requiring three to six years of service and a final evaluation based on their teaching, research, publication in journals of record, and community service. The tenure system tends to foster specialization at the expense of the interdisciplinarity and interconnectedness that are so essential to solving environmental problems. For this reason, Ball State University's "Green" Committee's Final Report recommends rethinking the existing traditional faculty tenure and promotion system in order to promote the interdisciplinary, short-term projects needed to make gains toward implementing the university environmental goals.[4]

Some faculty may be part-time lecturers; others may be on a term contract. Most have doctoral degrees and have completed extensive academic research. In general, they teach and conduct their research independently or with others from their own department or discipline, so the interdisciplinary nature of campus environmental action may not attract them. In some departments, faculty receive a portion of their salaries from externally funded research grants and are under intense and increasing pressure to generate research funds. Preparing research proposals can be a major drain on a faculty member's time.

The university's faculty have their own committee and decision-making structures, and often these structures do not include staff or students. Faculty committees oversee most aspects of university life and curriculum and are often active in defining and overseeing university environmental academic programs such as environmental studies and research on environmental problems, technology, or policies, but they are less likely to be involved directly in environmental stewardship. Faculty committees that have direct environmental impacts include the budget and planning committees.

**Staff** Universities and colleges employ a range of staff from highly trained and salaried professionals who run the libraries and manage the physical plant, to custodians and warehouse attendants paid by the hour.

Some staff jobs, such as human resources and accounting, are readily interchangeable with the business sector; others, such as fund raising and grant administration, are more likely found only in universities or other nonprofit organizations. Regardless of the job, most staff differ from faculty and administrators in that their jobs are usually more defined, their departments are often more hierarchical than the academic departments, and they are rewarded, evaluated, and paid in a system that is entirely separate from those of the faculty and most of the administration. Staff provide support to the academic and residential life but often do not have a direct impact on students' formal education.

The staff is often the element of the university community that undertakes significant and long-lasting environmental initiatives to reduce waste, improve conservation, and save (or cost) money. In particular, decisions with major environmental consequences tend to be made by departments such as the facilities management (buildings and grounds), materials management and purchasing, and dining. These departments manage large budgets, construct new buildings, heat and maintain existing facilities, buy materials, negotiate contracts, and prepare and clean up after millions of meals a year. Despite this potential to make a difference, many staff do not readily see the connection between their work and its environmental consequences. Similarly, many students and faculty seeking to improve their campus environmental footprint often fail to see the importance of staff involvement.

**Students** Students are the university's customers, and their education is the primary reason that most colleges and universities exist. Since the 1960s, students have gained increasing influence with university trustees, administrators, and faculty, yet they are often still held at arms length by these upper-level managers and staff employees. Students nevertheless may have more influence than they realize over the practices and policies of university staff, such as food services and housekeeping staff, who are generally very responsive to student interests and requests.

Students bring energy and creativity to solving campus environmental problems, finding resources, and collecting detailed information. They are eager and energetic and contribute a unique perspective to the process of environmental action, although they are often overlooked as a resource by staff and administrators. Students' own systems of governance can be effective vehicles for promoting environmental change. Students across the country have advocated for recycling programs and other environmental initiatives at many schools. Commonly, they provide the volunteer labor to begin and promote a program. The most successful and lasting programs are partnerships between students and university staff and administrators. At Tufts, student pressure and activism led to divestment from a utility company because its planned hydroelectric plant threatened sensitive ecosystems and indigenous people. A student lawsuit at UCLA forced improved ventilation in the art studios.

Student participation in campus environmental activities takes many forms, varies widely from year to year, and yields results that depend on the particular student group. Undergraduate and graduate students, residents and nonresidents, younger and older students, as well as environmental and campus leaders all respond differently to calls for action, volunteerism, and personal responsibility. Student turnover from graduation and annual matriculation results in both renewed energy and a lack of institutional memory. Each year students new to the campus or new to campus environmental issues are likely to raise some of the same issues that were evaluated and resolved the preceding year, a phenomenon that can be tiresome for staff who are called to respond.

At Tufts, students prefer to work on projects from their own agenda. On a number of occasions, Tufts CLEAN! staff asked the environmental student group to join in planning, promoting, or carrying out a project. The initial response was always positive and enthusiastic, but the outcome usually depended on the project's autonomy for students or with the way it fit with their own goals. Just as successful environmental stewardship led by students needs to seek staff input in the planning phases, staff, administrators, and faculty must also invite students into the planning and project development phases of their initiatives.

**Other Players** Certainly there are many other members of the university family: alumni, parents, funders, vendors, and the surrounding community. Comprehensive environmental initiatives can complement and use the skills of each of these groups. For example, college alumni or parents may be good sources of expertise about energy efficiency or alternative energy and may be willing to donate their services. The neighboring community may also be a resource, providing space, expertise, or partnership opportunities. Universities should be sensitive to neighbor concerns where environmental programs may pose a burden. For example, neighbors may bear the burden of increased traffic to a composting pile on the edge of campus, or object to the storage of recyclable materials in a parking lot near their homes. It is important that neighbor relationships be cooperative rather than adversarial if long-term trust and working relationships are to develop.

## Essential Ingredients for University Environmental Change

### University Commitment

Corporations such as Bristol-Myers Squibb, Dow Chemical, Dupont, and 3M have learned that top-level institutional commitment to environmental stewardship efforts is valuable to the success of their programs to reduce chemical waste and improve their environmental track records.[5] Top-level commitment is as important for universities as it is for corporations. University staff, faculty, and students often look to the university administration to articulate and follow through on commitments to environmental stewardship. At many schools, faculty and staff are skeptical about their administration's commitment to environmental initiatives and may even delay their own actions until they see top-level administrators demonstrate their intentions through actions, policies, and the investment of resources.

For environmental stewardship programs to be successful, the institution's leaders—the president, provost, vice presidents, and deans—must make a visible and meaningful commitment to environmental action. Sometimes this leadership begins with the top administrators; at other institutions, student, faculty and staff environmental leaders need to seek and cultivate this support. Ideally the first step in this commitment is a recognition by the administrators that:

- The university has varied environmental impacts.
- Reducing these impacts is in the long- and the short-term interest of the institution.
- The impacts go beyond the creation of trash, and thus the strategies must go beyond recycling.

- Some of the strategies save money, others reduce liability, and still others do neither but are nevertheless worth doing because they have long-term benefits for health, safety, the environment, or the community.
- Appropriate and creative solutions may require rethinking some current operating norms.

Demonstrated commitment to pollution prevention and natural resources protection requires both personal action and institutional support. In fact, once an administrator states that he or she is committed to environmental stewardship, everyone else tends to measure any actions against that standard. A failure by top-level administrators to assume personal action can stymie more comprehensive efforts and discourage participation. For example, a mailing from the president's office that is double-sided and printed on recycled paper is a clear symbol of consciousness and a model. Of course, administrators must take care that their visible commitment is more than symbolic. The goal is to ask of every decision what the environmental implications are and what can be done to reduce or eliminate adverse impacts.

Campus environmental advocates may find that they can build on this general commitment by offering their assistance to top administrators rather than by demanding action. Advocates, for example, can convene a fact-finding committee, make recommendations, and perhaps even draft letters for the president and the provost to review and distribute on official letterhead. In addition, advocates can provide opportunities for top administrators to demonstrate commitment with symbolic gestures such as tree plantings or an Earth Day visit to a department that has a model waste reduction program.

University or college administrators have the power to make the necessary commitment to environmental stewardship and to demonstrate it. Those other players outside the administration can also work to create the feeling and demonstration of commitment.

**Environmental Policy Statement**

This policy statement is a public declaration of university commitment to environmental protection and serves as the framework for decision making and goals. In addition, the development and subsequent announcement of a policy statement can be instrumental tools for raising issues of substance and developing consensus on policy. When a policy is absent or is developed without broad input, efforts are likely to be uncoordinated, and the result will be unfocused and short-lived. A policy statement is generally most effective if it is backed up by specific goals and implementation strategies. Specific policies that are well directed make implementation easier, since the responsibilities of end users are clear.

Many proactive corporations have well-established environment, health, and safety policy statements that articulate the corporation's position on environmental issues with varying levels of specificity. The chief executive officer or the board of directors usually endorses these policy statements. Universities, because their impacts are less understood and often less regulated, are much less likely to have a comparable policy statement or well-established mechanism for carrying out the directives provided in such a statement. Nevertheless, some universities have adopted broad environmental policy statements; among them are Tufts (see box 2.1), St. Lawrence University, and the University of Georgia.[6] Dartmouth's statement is short and simple but is backed by detailed objectives. Over two hundred presidents and rectors from thirty-five countries have signed the Talloires Declaration, a statement of principles that support environmental education and stewardship (see appendix A). Others have adopted specific directives for targeted issues as part of a general policy or as part of department- or issue-specific policies.

At Tufts, the process for developing the policy statement was nearly as important as the statement itself. In 1991, about twenty-five staff, faculty, deans, and students from throughout the university were appointed to a policy committee by the President's Council of vice presidents and deans. With a request for an environmental policy from President Jean Mayer and the knowledge that Tufts' policy would be the first of its kind at a university, the committee studied corporate environmental policy statements. Committee members submitted written drafts of policy statements individually, and Tufts CLEAN! staff and a student volunteer condensed similar policy language from each of the dozen individual policies into a large draft. The committee reconvened and examined the draft, selecting and crafting the language from each section to reflect their intent. The final document was submitted to the deans and vice presidents for ap-

**Box 2.1**
Tufts University environmental policy

We, the Tufts University community, affirm our belief that university faculty, staff, and students have a responsibility to take a leadership role in conducting activities as responsible stewards of the physical environment and using educational activities to promote environmental awareness, local action, and global thinking.

In our university functions, Tufts University will strive to:

- Conserve natural resources and support their sustainable use.
- Conduct affairs in a manner that safeguards the environmental health and safety of students, faculty, staff and communities.
- Reduce the use of toxic substances and the generation of wastes and promote strategies to reuse and recycle those wastes that cannot be avoided.
- Purchase renewable, reusable, recyclable, and recycled materials.

In our education and research missions, Tufts University will strive to:

- Foster an understanding of and a responsibility for the physical environment.
- Ensure that individuals are knowledgeable about the environmental and health issues that affect their discipline.
- Encourage environmental research.
- Conduct research and teaching in an environmentally responsible way.
- Provide a forum for the open flow of information among governments, international organizations, industry, and academia to discuss and study environmental issues and their relationship to other social issues.

In our student and employee relations, Tufts University will strive to:

- Delineate individual responsibility and guide action for ensuring safety and minimizing adverse environmental impacts in the implementation of this policy.

Tufts will consider full compliance with the law to be the minimally acceptable standard and will exercise whatever control is reasonable and necessary to avoid harm to public health and the environment, whether or not such control is required by regulations.

Tufts will initiate, promote, and conduct programs that fully implement this policy throughout the university and the global community.

proval and was signed by the president on Earth Day 1991. When Tufts' new president, John Di Biaggio, joined the university in 1993, representatives for the environmental committee met with him to introduce the policy. At that time he asked the committee to review it. One minor modification was made, and he signed the policy soon after.

University policies and the subsequent implementation of those policies in the departments of food services, purchasing, and facilities management offer powerful opportunities to implement and institutionalize environmental stewardship efforts. Policy statements that are department or issue specific can be more effective than general university-wide policies. For example, some institutional purchasing departments have a great deal of latitude to make or carry out policies that require the purchase of recycled paper, the use of designated vendors, or requirements by vendors of some products to take back spent product or packaging. Tufts dining services developed a detailed statement of their environmental actions as part of their strategic planning process (see chapter 5). The University of Buffalo has adopted a series of issue-specific policies covering issues including air conditioning, third class bulk mail, campus newspapers, recycling, and purchasing. Even the U.S. government, through presidential executive orders, mandates the procurement of recycled paper, Energy Star (energy-efficient) personal computers, and alternatively fueled vehicles for many government applications and government contractors.[7]

### Environmental Committees

Many people in a university have environmental responsibilities yet may not fully understand these obligations or see how they are related to their jobs or to others at the university. A university-wide committee can help institutionalize environmental stewardship efforts and bring stakeholders to the table. It may oversee broad-based environmental action or take a strong hand in shaping detailed plans of action. The committee membership should reflect the university community (students, faculty, staff, and administrators).

The committee should be sanctioned by top university administrators. If this commitment is unavailable, for whatever reason, organizers should nevertheless charge ahead. If the administration can provide a specific and progressive mission or task for the committee, such as policy development or meeting energy reduction goals, the committee's job is easier. For example, West Chester University's president asked the environmental committee to develop a plan for environmental improvement on campus.[8]

Many universities are experimenting with university-wide committees or smaller issue-specific committees. Whatever the structure, a well-run committee can be effective for securing ideas, motivating, and rewarding change. The committee may also provide an open forum for the university community to raise issues and learn about ongoing actions. Each university or college will find that the makeup and mission of its committee is unique, but the most successful committees draw on interested and willing members of the community and try to gain wide representation from staff, faculty, and students.

Forming the environmental committee may be easy; leading and directing it so that its meetings are productive is more difficult. Committees that are most successful discuss policies that affect many departments and assign detailed implementation of projects and programs to subcommittees. The committee can also be effective in bridging communication gaps among faculty, staff, and students or for advocating for new ways of thinking across the university. Information exchange can be effective if committee members carry the information they learn back to their departments. Tufts established the University Environmental Improvement Committee to oversee the development of the Tufts environmental policy statement and then to focus on its implementation. Tufts CLEAN! staff identified the need for the committee and suggested that the university president ask the deans to designate representatives to this committee. It was somewhat surprising that the Tufts Environmental Improvement Committee operated effectively for nearly four years, despite the fact that it was ineffective at addressing many concrete and specific actions after the policy was developed. Committee members continued to come to regular bimonthly meetings during the school year, several traveling from other campuses as much as forty miles away, perhaps because they felt that the top administrators were committed to the issue. Furthermore, committee members repeatedly told Tufts CLEAN! staff that one of the most valuable parts of each meeting was the members' updating each other on their environmental actions or problems to date. This discussion provided them with opportunities to share their successes, learn from each other, identify common problems, and feel that their efforts were important. In addition, it provided staff, like the director of community relations, with opportunities to identify potential problems before they became actual problems. The committee as well provided valuable feedback to Tufts CLEAN! and other environmental groups throughout the university. We debated mandates for recycled paper, energy-saving initiatives, divestment from polluting companies, carpooling, and safety concerns.

The initial leadership of Tufts Environmental Improvement Committee was provided by an academic with lots of good ideas and enthusiasm. The committee was later and most effectively chaired by the director of physical plant, a good manager who called himself a "reformed environmental abuser." He fully understood how the university works, saw the connections between operations and environmental stewardship, and was able to conduct a well-run meeting.

In all cases, Tufts CLEAN! staff worked behind the scenes to prepare agendas, background material, meeting minutes, and desired meeting outcomes. The Tufts committee ceased to meet when funding for Tufts CLEAN! ceased, an event nearly coincident with the departure of the committee's chairperson. However, in a recent "earth summit," students, faculty, operations staff, and administrators who gathered recommended that Tufts reconvene a university-wide environmental committee, which they saw as the best way to support ongoing environmental efforts and foster support for new initiatives.

Like Tufts, Brown University launched its campus-wide program (Brown Is Green) with a large university-wide committee, chaired by the provost. Now that Brown Is Green is well established, the larger committee rarely meets and instead has subcommittees that work on the various objectives. Among the other campuses that have university-wide committees are Ball State University, St. Lawrence University, the University of Vermont, the University of Buffalo, and Bates College.

Several schools have used specific waste reduction or solid waste committees. Subcommittees of the University of New Hampshire's Committee on Energy Conservation and the Environment include groups on building systems, education, bicycling, precycling (using materials a second time prior to recycling), and electric vehicles. MIT has a committee that works to develop strategies to reduce, reuse, and recycle hazardous materials; its members include the vice president; environment, health, and safety staff; and purchasing agents.[9] Many schools have waste reduction, solid waste, or recycling committees to help address the logistical and educational needs of solid waste and recycling issues.

At Tufts, we found that safety issues, especially those that focus on chemical safety, are often best dealt with outside the large campus-wide environmental committee. Safety committees on the Tufts medical school campus and the main campus included participation from faculty and teaching assistants from departments with laboratories, art studios, buildings and grounds, and others who regularly use hazardous materials (see chapter 7). The environmental, health, and safety staff play a large role in these committees.

**Individual Leaders**

Most successful environmental efforts on college and university campuses rely on leaders throughout the institution. While the university community's awareness of and motivation toward acting on environmental problems ebbs and flows, the catalysts for the essential broad-based action almost always stem from one or several dedicated leaders in the effort. These leaders of university-wide action can come from anywhere within the university, from the level of the provost to the faculty and students, or even a member of the staff. In all cases of university action, individual leaders play an essential role in making a difference.[10] Buy-in and leadership from a variety of levels within the university are necessary because many environmental initiatives must be integrated into the everyday operation of the university. In the green university, administrators, faculty, staff, and students take responsibility for their own actions and have an understanding of the ways that their actions affect the environment. They are informed by environmental leaders: people who spend time teaching others, investigating opportunities, and executing projects.

At many universities, individuals take on initiatives that are outside their job responsibilities to demonstrate commitment to conservation, waste reduction, and recycling. For example, at the Tufts medical school, an administrative assistant took her environmental commitment so seriously that she ordered and used recycled paper in the entire department for two months before letting anyone know, thereby anticipating and diffusing the department chair's resistance to the use of recycled paper. She used her knowledge of the department's decision-making structure to inform her environmental stewardship action. At Rutgers University, the senior buyer, Kevin Lyons, went well beyond the strict definition of his job responsibilities and initiated extensive procurement standards for the university, resulting in a dramatic increase of recycled products and an improved recycling program.

Key faculty members can lead the campus environmental efforts. David Orr at Oberlin College is an outspoken leader for innovative thinking and action, on his own campus and on other campuses across the country. Anthony Cortese, former director of the Center for Environmental Management and dean for environmental programs at Tufts, was influential in helping Tufts' president to embrace environmental education and stewardship as leadership opportunities for the university. Harold Ward, at Brown University, and other environmental studies directors across the country are instrumental in linking campus-based environmental action with environmental studies curricula.

Students can also be leaders for campus action. There are many examples of student-led initiatives that have reduced waste, expanded recycling programs, and developed innovative and far-reaching solutions to campus environmental problems.

**Campus Jobs with Environmental Leadership Responsibilities**

Only a few schools have established an environmental leader position, or environmental ombudsman, with the responsibility for researching, catalyzing, and influencing environmental stewardship action on campus. An environmental ombudsman can be dedicated to addressing unregulated impacts, serve as a resource, and provide a link between the academic and the operational parts of the university. The ombudsman should have both academic and operational responsibilities—rare in most universities—to articulate the commitment needed, influence the financial and personnel resources, promote efforts, and assist where needed. An ombudsman can serve as a catalyst for action and provide assistance to others throughout the university who are taking action.

The environmental ombudsman's position is probably the best model for leading and achieving far-reaching campus stewardship. If the individual is well respected and diligent, he or she can influence and support academic and operational policies. However, few universities have been successful in maintaining this position as a dedicated, university-supported, long-term position.

Until 1995, the University of Kansas's environmental ombudsman reported directly to the provost and provided guidance and assistance on issues ranging from the use of ozone-depleting chemicals to the purchase of recycled products. The success of that office, evident in the number of actions it influenced and the breadth of its effectiveness, was probably related to degree of the commitment from the administration and the personality of the ombudsman and his staff. Tufts CLEAN! was funded by a grant from the Environmental Protection Agency and was located in a research center well outside the university operations. A small staff and graduate students identified and studied problems and came up with solutions, led the university environmental committee, and provided assistance to other members of the university interested in environmental action. When outside funding was no longer available, however, the office could not be sustained. At Brown University, the Brown Is Green ombudsman's role is an outgrowth of an academic position to teach a course using campus-based student projects. The position was originally funded by the environmental studies department and the operations departments. Now the position is funded entirely by the academic side of the university, although the ombudsman continues to pursue many operational issues. In contrast, at the University of Wisconsin, Daniel Einstein, the environmental management coordinator (officially an administrative program manager), is funded by the physical plant department despite his large role working with students on academic projects.

Increasingly universities are hiring recycling coordinators or energy managers, some of whom have assumed personal responsibility for spearheading environmental action campus-wide. These positions are extremely valuable, but their existence does not substitute for members of the entire university community taking responsibility for their part in the problems and their role in the solutions. Recycling and energy managers should be expected to help motivate and lead others; as well, the recycling coordinator, the environmental health and safety staff, and the energy manager are charged with specific responsibilities—for example, the collection and marketing of recyclable materials, the disposal of chemical wastes, and the efficient use of energy. Because these responsibilities have environmental as well as regulatory and financial benefits, the university tends to defer to the people who fill these positions for all environmental issues.

## Environmental Protection Principles

Keeping the principles of energy efficiency, waste reduction, safety on and off site, and long-term and comprehensive (or life-cycle) cost accounting as overarching objectives can guide most decision making to conclusions with lower environmental impacts. This understanding of environmental protection principles then empowers members of the university community to make decisions. For example, staff members in the purchasing department can evaluate similar products and choose the one that best reduces waste or improves safety without needing specific guidance on each and every product. These principles inform decision making that must occur on the product-by-product and decision-by-decision basis.

### Energy Efficiency

Using energy (electricity, oil, natural gas, etc.) only when needed or with the most efficient technology possible helps to reduce the impacts on air quality, resource depletion, global warming, and acid rain from burning fossil fuels for heat, cooling, hot water, and electricity. Energy-efficiency practices can save money in all climates. Efficiency practices should be applied to new construction as well as small- and large-scale renovations and to electrical equipment (e.g., motors) and lights, as well as heating and cooling systems. Even when fuel costs are low, efficiency is a good investment for the future, for energy costs are likely to rise. Implementing energy efficiency on campuses usually requires careful attention to existing conditions and systems and a knowledge of new technologies. Nevertheless, there are many low-tech opportunities to improve efficiency through behavior change, such as turning off lights and equipment, and institutional policy, such as temperature targets and heating and cooling policies.

### Waste Reduction

Waste reduction is the principle of efficiency applied to other materials, such as packaging, chemicals, food, and water, and can reduce the generation of solid and hazardous waste, as well as the resources and pollution from the manufacture and transport of goods. Successful waste reduction is often incremental; as long as any waste exists, there is room for improvement.

People most often think of waste reduction with regard to hazardous waste, such as reducing the generation of hazardous wastes (e.g., paint thinner and used oil), but the principle of using only what is needed and avoiding spills or other waste translates to water use, fuel use and storage, packaging, and paper products such as copy paper and paper towels. Successful waste reduction requires a combination of technology (such as the substitution of less toxic chemicals), policy (including prohibitions in the purchasing of certain chemicals or materials), and individual behavior change in everyday practices.

### Safety

Safety is an important and familiar principle that usually dovetails with environmental protection objectives. On-site safety and environmental issues coincide on issues such as the handling, use, and disposal of paint thinner, pesticides, and concentrated cleaning or degreasing products. Chemical, asbestos, and PCB handling have safety consideration for workers and building occupants. Off-site impacts are less visible than on-campus impacts but nevertheless should be considered when making decisions. It is important for campus environmental advocates to combine safety with environmental considerations; safety is compelling and usually has benefits for environmental protection as well.

**Life-Cycle Costs**

Life-cycle costs refer to the costs of a product from its manufacture to its disposal. Outside environmental circles, people may think of this same principle as "thinking long term." When a product's life cycle is considered, often the lowest initial cost is not the lowest overall cost. For example, a small, unreliable copy machine may not be as cost-effective over its lifetime as a larger, energy-saving, double-sided model even though its purchase price is lower. The second machine may waste less paper and use less electricity. Life-cycle costing also considers off-site effects in the product's manufacture and disposal as waste even if these effects are not evident to the product's users. Information about off-site impacts is often difficult to find, but university decision makers can work with student researchers to investigate these questions. On-site but harder-to-quantify impacts, such as hazard to workers using a cleaning product or the danger of a spill from a chemical, should also be considered in life-cycle decision making.

Using life-cycle analysis tools is a useful exercise, although it can be time-consuming and sometimes academic. Most environmental decisions need to be made more quickly than a complete life-cycle analysis allows, but the basic principles can still be applied to university decision making. One difficulty in thinking about life-cycle costs is that it often requires comparing unlike impacts—for example, the impacts to air as compared to those on the water. The Center for Regenerative Studies at the University of California at Davis used some life-cycle analysis in the design of its innovative new dormitory building. The members selected a copper roof because it has a long life despite the energy-intensive nature of its manufacturing process—a trade-off that is difficult to quantify and compare even when there was a commitment to evaluate the available quantitative measures.

## Gathering and Using Data

In order to make campus environmental change happen, targeted and detailed data of several types must be gathered. These data will inform change, provide a baseline for measuring progress, and allow realistic solutions to emerge. Despite this need for data, many who are undertaking campus environmental stewardship are collecting the wrong types of data, collecting data without understanding how they will help make change happen, or spending more time collecting data than implementing programs. The academic climate of universities often allows the collection of data for the sake of data, sometimes at the expense of change.

Data are important for informing decisions and evaluating solutions. They are needed to make decisions about the outlay of capital or to change a waste disposal method. Nevertheless, some solutions to university environmental problems depend little on the magnitude of the problem. For example, increased carpooling and more bike racks will be part of the transportation solution in any place where people drive to and/or live close enough to bike to work. The total electricity or water used does not change the solutions needed to use them more efficiently, especially when a program is starting. To make truly informed decisions, decision makers need to know in detail the nature of existing technology, available solutions, and the infrastructure (e.g., the lighting and the locations of older toilets and inefficient shower heads, available replacements, and the condition of the plumbing or building wiring). A broad-based audit will be useless if it neither compares the track record of the college or university to other similar institutions nor provides the detailed information needed to make change.

Data gathering should be part of the process of environmental improvement rather than an end unto itself. The key to collecting the right data is to be clear about the project goals and match data collection to that objective. Many people gather far more data than is useful simply because the information is there. Students and other environmental advocates must realize that data gathering puts demands on others, especially buildings and grounds, purchasing, dining services, and accounting staff members, who will be essential when the time comes to carry out change.

**Environmental Audits**

Across the United States, students and others are undertaking broad-based environmental audits of their college and university campuses, quantifying the university's environmental impact by gathering information about resources used and wastes produced by the university and collecting general information about opportunities to reduce or eliminate those impacts. Environmental audits are valuable for raising awareness of university environmental impacts. For example, aggregate data that put annual demand for electricity in terms of the number of days that the nearby nuclear power plant must operate can attract attention to electricity use and its consequences. The environmental audit can be labor intensive as students collect mountains of data from the departments of facilities, food service, purchasing, safety, and community relations.

The Tufts campus environmental audit found that annually we threw out 65 tons of used paper towels, burned 1.1 million gallons of heating oil, used 110 million gallons of water, and produced over 2,000 tons of solid waste. Determining these figures was not easy. Many of Tufts' purchasing, utility, and water records were incomplete, not computerized, or in a form not suited to the audit. Tufts staff were sometimes hesitant to provide the information because compiling it took time, data provided for some previous project had been misinterpreted, and the purpose of the data collection effort was unclear. After nearly a year gathering general and descriptive data, the Tufts CLEAN! team finally concluded that a broad-based audit was of little use in informing action because we had not gathered the information that we really needed. For example, instead of detailed data about the specific use of resources and technologies in a building or even a room, we had gathered aggregate information about total energy and water use. In addition, despite many interviews, we had not asked university staff to help identify opportunities, relying instead on environmental literature and general concepts. Furthermore, Tufts concluded that the broad-based campus audit could actually be detrimental to environmental action because of the general nature and heavy burden the audit places on university administrators whose time is better channeled into implementation of environmental actions.

Detailed audits of particular problems, however, are essential for measuring progress, informing decisions, and evaluating the project in environmental terms, such as pounds of waste reduced or kilowatt-hours of electricity saved, as well as to measure financial savings. Measurement schemes can vary from basic program-oriented measures (e.g., the existence of a program), to activity-based measures (e.g., specific processes or operating activities), to quantitative measures (e.g., tons of solid waste as compared to a baseline).[11]

The appropriate measurement scheme to use in a given decision at the university depends on measurement objectives—to motivate individuals, justify expenditures, create publicity, or determine effectiveness of action. Corporations' measurement schemes depend on the process and the industry.[12] In a university, numerous processes result in a wide range of environmental impacts, and although data may be plentiful, they are often too poorly organized to measure progress. Broad campus audits provide a quantitative and often aggregate snapshot of environmental impacts but may not be suited to measuring pollution prevention progress.

The key to an audit is to gather appropriate measures. Data should be used to inform a decision or determine progress. For example, a measure of total electricity use does not tell how to reduce that use or if progress has been made. Instead, components of that electricity use (e.g., from lighting in one building, from computer use, or from a specific facility) are useful for determining success.

### Data for Measuring Progress

Progress on campus environmental initiatives can be measured in a variety of ways ranging from simple to sophisticated. To measure and promote progress requires attention to the data that are available and the ways to make sense of them. Quantitative, qualitative, purchasing, and normalizing data can all be useful in measuring progress (see box 2.2). For example, an effort to measure the success of a program to curb the generation of solid waste in residence halls should determine if there is a way to collect data on the relevant portion of the waste stream as distinct from the data on the waste stream as a whole. Furthermore, if the data are available by weight, it would be foolish to measure the progress of the program by volume. The effort to measure progress should also identify factors that confound a program, such as changes in hours of operation or special events.

**Quantitative Data** Measuring progress quantitatively is the most appealing measurement scheme. Knowing how much waste was reduced or the dollars saved is appealing, and the figures are readily understandable.

**Box 2.2**
Data to measure progress

**Quantitative data**
Measured change on net
Measured change from specific action
Utility data (electricity, water, oil, solid waste)

**Qualitative data**
Lists of actions

**Purchasing data**
Paper
Food
Chemicals
Equipment
Contracts (e.g., pesticides, fertilizers)

**Normalization data**
Students
Faculty and staff
Square feet
Acreage

However, many environmental actions cannot be accurately quantified—for example, the reduced risk of accident or the benefits of reduced air pollution. Furthermore, to measure change accurately, the data must reflect the results of the action alone with all other factors constant. For example, electrical efficiency gains may be achieved through lighting retrofits in many projects, but the savings do not keep pace with the increasing electricity demand from new computers and equipment. Often it is more effective to measure the magnitude of change from the specific action or project itself. For example, one department's efforts to reduce the generation of waste will be imperceivable in the campus-wide solid waste tally despite dramatic reductions in the department's own solid waste.

Data on a college's use of utilities are important quantitative measures for environmental stewardship programs because the use of electricity, oil, natural gas, and water and the generation of solid waste have direct environmental and financial consequences. All institutions collect utility data in some form for billing and payment purposes; however, the billing files may not be useful for determining the success or failure of an environmental program. Aggregate utility data are difficult to use in assessing the success of many initiatives, because the changes that result from direct action may be only a small portion of the total bill. In fact, the successes of some efforts will be totally unnoticed if they are measured only in their effect on total electricity, oil, or water use.

Measurements of an institution's use of water, electricity, and heating fuels offer a snapshot of the vast amounts of resources that a single entity uses. Analysis of the relative magnitudes of the use and cost by year and by building type gives a sense of trends, successes, and areas for improvement. A detailed look at utility data can also help identify billing errors or problems with building systems. Clearly an analysis is only as good as the numbers and data inaccuracies are commonplace, but the net totals are still useful for spotting unusual patterns. For example, monthly examination of the water meter can help identify large leaks, and comparing the electricity used per square foot of two buildings with similar uses can identify irregularities or problems with overuse of equipment. Revising utility records in a way that will help future analyses can be very useful for comprehensive environmental efforts.

**Qualitative Data** Qualitative data are lists of actions that have been undertaken to reduce waste or improve conservation without specific quantitative measurements of their results. Examples of these qualitative measures might be the establishment of a waste reduction committee, the adoption of an environmental policy, or the addition of language requiring the recycling of waste paint in a painter's contract. The Environmental Program at the Park Plaza Hotel in Boston gained a great deal of publicity for its more than sixty-five environmental initiatives. Although detailed analyses and quantitative background are available for some of these, others were publicized and listed simply as action taken. At Tufts, we found that publicizing and praising the existence of an effort or initiative, even it we could not quantify it, was a powerful motivator for the actions of others and the continued commitment of the staff who had taken the initiative. The actions discussed in detail in chapters 3 through 9 could be used as a list of qualitative actions for a university to embrace in its future environmental stewardship program.

**Purchasing Data** Data about the amount of material purchased by the university can be a useful measure of environmental change and the effectiveness of environmental stewardship programs. Data about goods and services purchased can provide quantitative and qualitative information to measure progress and inform change. When the university purchasing department provides a centralized purchasing function, the department records hold information about quantities of material used. Thus, the department can be a valuable source of information about the characteristics of the products—the recycled content of paper, for example, or the gas mileage of university vehicles. In most institutions, the purchasing department also negotiates contracts for items and services such as double-sided copy machines or the disposal of waste motor oil.

Purchasing records are not always designed for analysis by item and may simply be a file folder of purchase order slips rather than a computerized database that is ready for analysis. In some schools purchasing practices are highly decentralized, so data will be nearly impossible to assemble and evaluate. As with other data, it is important to determine the purpose of data gathering before taking time to put the data in a form that is usable for analysis.

**Normalization Data** Data used to put other data in forms that can be fairly compared are called *normalization data*. To compare departments or buildings, or the success of one recycling program to that at another school, it is important to measure progress by a metric that accounts for the size of the program or school or the change in size over time.

Many universities have a good accounting of the size of each building (in square feet), so size may be a useful metric for normalizing information on energy use by building for comparison. To normalize some data, it may be most appropriate to use annual graduates, since they are a university's "product" and the reason for a university's existence. However, measuring progress by the number of graduates or enrolled students can be problematic since this measure does not account for the research that a university may conduct since the quantity of research is not usually a function of the number of students. Nonetheless, comparative analyses—between schools, departments, or even over time—should be

**Table 2.1**
How normalizing data changes the conclusions

| | University A: Private, Ivy League | University B: Public state school |
|---|---|---|
| Total electricity use | 100 million kWh | 50 million kWh |
| Total square feet of space | 50,000 square feet | 10,000 square feet |
| Total student population (full-time equivalents) | 5,000 students | 10,000 students |
| Electricity per square foot | 2,000 kWh per square foot | 5,000 kWh per square foot |
| Electricity per student | 20,000 kWh per student | 5,000 kWh per student |

measured in some way to ensure accuracy. The choice of this normalization scheme can be important in learning the outcome of the analysis. Table 2.1 provides a hypothetical example of how normalized comparisons between schools might show that an Ivy League school uses more energy and water resources per student than a community college because the more prestigious school has more facilities for each student and tends to use its facilities less intensively (fewer evening and weekend classes, for example), even if these facilities are more efficient on a per unit (square foot) basis.

## Data for Informing Change

Data for informing change are essential and must accompany every action. Box 2.3 summarizes examples of these data. Successful environmental programs reflect the reality of the university: how it works, what its goals are, and how the existing constraints may limit action, including the physical, financial, and human constraints met by the existing equipment or procedures. Faculty and student environmental audits often overlook data that inform change, such as data about the details for a project, the attitudes of those who must undertake or oversee a project, and anecdotal data, from which much can be learned.

Sometimes students and academics tend to favor quantitative data at the expense of data that inform change and are essential to implementing environmental action. The Tufts CLEAN! research staff spent countless

**Box 2.3**
Data to inform change

**Detailed data**
Lighting: lamps, ballasts, switches
Heat: thermostats, radiators, distribution
Hours of use
Hours of need

**Attitudinal data**
Perceived needs
Willingness to change
Surveys
Evaluation

**Anecdotal data**
Why past efforts have failed
Historical information
Opportunities

hours understanding and analyzing data on the historical use of electricity. What we needed instead were detailed data about each lighting fixture in targeted buildings. The goal of recycling cardboard from dining services was more successful; we relied on crude estimates of cardboard generation, made by collecting all cardboard for a single day, in order to determine the amount of storage and the size of the baler unit we needed. We spent the most time meeting with unit managers, loading dock staff, kitchen personnel, and truck drivers to understand what would and would not work. Cardboard generation, consolidation, storage, and hauling to the baler were all worked out with the staff. In that effort, the quantitative data we did collect—the measurements of the baler unit bought to consolidate the material—failed us. When the seven-foot-high unit was delivered to the loading dock with an eight-foot ceiling, we discovered it had a piston that extended to ten feet high. A hole in the ceiling solved the problem.

Gathering the data needed to inform change is less systematic than compiling data that measure progress. To gather these data, we must find answers to a number of questions:

- How are things done now, and why?
- Where does waste occur?

- How can that waste be reduced or eliminated?
- Has a potential solution been tried before?

**Detailed Data** Data about the details of a process or practice are often the most useful information for informing change, conducting cost-benefit analyses, or determining actual costs. For example, improving the efficiency of campus lighting can save money and avoid the emissions that result from the generation of electricity. Knowing the detailed specifications of the existing and desired light levels, the lamps, the ballasts, and reflectors are necessary for success. Without this information, informed and cost-effective decisions cannot be made.

One of the most common mistakes that students make as they audit their campuses for environmental improvement opportunities is to assume that aggregate information can inform successful decisions, so they often overlook the need for detailed information. Students who are trained in the specific data collection techniques needed for a particular project (perhaps in a for-credit internship) can be a cost-effective resource for gathering detailed program information. They can be particularly effective in counting equipment such as computers, incandescent lights, flush toilets, and electric ovens. They can also make many detailed observations and identify needs and constraints. For example, to learn how to improve systems, students might ride the recycling truck to see where parking or access limits collection or where curb cuts are inadequate for the dollies. The Nebraska State Energy Office has trained student auditors to perform in-depth energy audits in public buildings, schools, and universities. This combination of expertise through training and close supervision and student enthusiasm is a good model for successful student audits.

Information about the way that people use buildings, computers, ovens, equipment, and other resources is invaluable for designing programs and initiatives. Often a simple reduction in use, such as turning off equipment when it is not in use, can decrease the demand for resources and save money. A single Tufts dining hall discovered it could save $1,000 annually by shutting off equipment or turning it on only as needed. To determine these opportunities, we talked with people about what they needed to use equipment for and compare that information to how the equipment was actually used. Asking the people who use equipment *and* observing on a spot-check basis is most effective. Often the results of these anecdotal and observed data collection methods vary widely.

**Attitudinal Data** Information about perceived needs, willingness to change, and barriers to change are important for designing realistic programs. Attitudinal data are invaluable too for assessing and revising a program after it is up and running.

To gather attitudinal data, ask the affected people directly. For example, one might ask the safety officer responsible for traffic and parking violations about the best way to enforce a policy of preferential parking for carpoolers. Written surveys can be helpful in gathering information about general receptiveness to change, such as the willingness to carpool, collect recyclables, or share magazine subscriptions. Surveys also serve the purpose of conveying information to the respondent and can be useful as a communications tool by asking questions such as, "Did you know that our university has a carpooling program?"

Surveys can sometimes be unreliable and difficult to analyze or collect in-depth or anecdotal data. But universities are rich with expertise in designing and analyzing information from surveys, and a survey can be a wonderful student project.

Making a distinction between people's needs (e.g., the need to have adequate light or the need to arrive at work in a timely fashion) and the means of achieving these needs (e.g., energy-efficient lighting or carpooling) is important when collecting information about attitudes. Many cling to the status quo because they believe that it is the only way to meet their needs. Successful programs meet the needs of the community but may change the means of meeting the needs.

**Anecdotal and Observation Data** Anecdotes and observations about the way things work and why or about past efforts to address problems are often undervalued, particularly in an academic environment that values quantitative data. But for those who are concerned about a university's green policy, these anecdotes are invaluable for gathering information about the way a university has historically interacted with its environment and for identifying opportunities to improve—opportunities that are often overlooked by the very individuals who are striving to create a better policy but are relying too heavily on quantitative data. Anecdotes helped Tufts CLEAN! to learn why a 1970s carpooling program failed and why neighbors were likely to object to an on-campus compost pile. Anecdotes teach a great deal about the practical (and impractical) reasons for the existing way of doing business.

Far too often, quantitative data gathering is more detailed and time-consuming than it needs to be. Simple observation may be sufficient for gathering the information that is needed to make a change. For example, many environmental efforts begin with "dumpster diving": efforts to sort, categorize, and quantify trash in the dumpsters. A truly thorough analysis of a dumpster's trash would examine the trash at several different times of the school year and on different days of the week (a dormitory dumpster will have more pizza boxes and less paper on the weekends). An often overlooked trick for gathering data about solid waste generation or information about the presence recylables in the trash and materials that frequently go to waste is to ask the people who frequently look in the dumpsters—the custodians—or to look at the dumpster and observe the relative quantities of materials. Records of complaints about rooms or buildings that are too hot or too cold, often recorded by buildings and grounds departments, are another example of an easy way to gather data and pinpoint areas where conservation can occur or where improvements in thermostat control, thermopane windows, or insulation are needed most. By looking around for quick and easy ways to assess program needs, environmental stewards can spend less time collecting data and more time designing and carrying out programs that work.

**Data About Products and Technologies**

Information and data about specific products, services, and technologies that are in use or installed at the university, or could be installed, are important for understanding the university's impacts on the environment and the opportunities to change those impacts. Information about the size, cost, and performance of new equipment is important. Which equipment will fit the space? What will be effective at solving the problems and reducing impacts? How durable will the product be? These are only some of the questions that need to be answered in purchasing decisions.

In environmental decision making, the environmental consequences or the environmental performance of a product may confound the issues. Furthermore, the needed information may be incomplete, unavailable, or unreliable. For example, little, if any, regulation guides the labeling of products as recycled, biodegradable, nontoxic, or environmentally friendly, so data gatherers need to scrutinize the ingredients or manufacturing processes whenever possible in order to assess their environmental impacts. In the end, decision makers will have to settle for incomplete information and realize that new science or engineering in the future may change decisions made today. Nonetheless, the business of environmental stewardship is one of continuous learning and improvement.

### Data Gathering and Action

Regardless of the accuracy and amount of data collected, data collection alone is insufficient. Rather, action on this information is the most important step for environmental change. This advice may seem self-evident; yet consider the bookshelves that are full of thorough academic studies with good ideas that have never come to fruition. Implementing projects requires the direct involvement of the affected people, often university staff, in data collection and the development and testing of recommendations.

## Making Environmental Change Happen

There is no formula for making environmental change happen, and there never will be one. Hard work, dedication, and money do not ensure that change will happen or last, although they may be the most important ingredients. The seemingly slow pace of change can be especially frustrating for students who are on campus for only thirty weeks a year for four years. Most often, in fact, change happens in surprising ways. Several addition factors are helpful in greening the university.

### The Economics of Environmental Stewardship

Like it or not, the language of the world is money and thus we must often communicate our efforts to green the university in financial terms. It is nearly impossible to assign a dollar value to the benefits of clean air, potable water, and open space, but, happily, many of the most important environmental initiatives can have real financial benefits that are usually related to cost avoidance or avoided liability.

Universities and colleges receive their operating funds from four major sources: tuition and fees, research funding from government or foundation grants, donations, and interest on the endowment. Major expenses are salaries, operational expenses (utilities, trash removal, grounds and maintenance, etc.), capital expenses, such as the investment in a new boiler, and equipment. Universities and colleges vary widely in the extent to which they rely on tuition and can augment their operating expenses with income from the endowment or outside research grants.

College and university business and financial officers understand both long-term and short-term cost avoidance and the benefits of avoiding costly fines and potential liability problems. It is important for environmental leaders to quantify the financial benefits and costs of a new environmental technology, such as new photocopiers, or of a new policy, such as a mandate for the purchase of recycled paper campus-wide. It is usually better to complete a project for less money (or greater savings) than originally predicted than the other way around. Project costs must include the cost of existing and additional labor (salary and benefits). Operational costs, such as electricity, waste, and water, are important in order to educate the university that these are in fact costs of many decisions (such as building a new laboratory) that are often overlooked.

**Finding the Biggest Opportunities** The opportunities for financial savings from environmental stewardship activities differ from one campus to another, but on most campuses improving energy efficiency will offer a dramatic return on the university's investment, as well as decrease on- and off-site pollution. For example, in regions where electric power is more expensive (seven cents per kilowatt-hour and up), installing new technologies to improve the efficiency of lighting and motors can have payback periods of as little as three months; and more comprehensive projects that combine the installation of rapid and slower payback technologies will pay back in three to five years. Some electric utility companies offer rebates to subsidize efficiency measures, such as the installation of efficient technologies, dramatically reducing the costs to the institution.

(Rebates are cost-effective for the power companies because the new technologies reduce demand for electricity, effectively gaining capacity for additional electricity users more cheaply than by building additional generating capacity.) Reducing liability from chemical accident, hazardous waste spills, and oil storage can also save a great deal of money in avoided fines, suits, and liability. Although most schools begin their environmental initiatives with recycling efforts, solid waste programs will not accrue major savings unless they can succeed in reducing waste generation in the first place.

**Long-Term and Comprehensive Accounting** Since most colleges, universities, and schools plan to be in business in ten, twenty, or even one hundred years, it pays them to think in the long term whenever possible. Certainly most facilities planning has two-, five-, and ten-year horizons, but often the payback periods must match those of more traditional investments in financial markets, stocks, and bonds. Often overlooked is the cost of redoing work or delaying work until the future, when materials are more costly, the cost of fuel may be higher, and the availability of capital scarce. Without a comprehensive approach to accounting and budgeting, the savings from many projects cannot be realized. This problem is most evident in construction and renovations that fail to take advantage of the newest technology, such as energy management systems or materials with good insulating properties in favor of lower construction costs but higher maintenance costs.

**Full Costs to the University** The costs of electricity and the generation of solid waste in a special project or new facility are often overlooked. For example, when a department builds a new computer room or electronics laboratory, the cost of the increased electricity demand rarely figures into the costs of construction or facility operation. Furthermore, many campuses have centralized or antiquated meters for assessing the electricity demand in a building or school, and therefore they may not actually be able to see these increased costs. When university administrators and managers fail to look comprehensively at overhead costs, such as waste disposal, heat, water, and solid waste handling and disposal, and apportion these costs accurately to the departments that use them, they overlook a portion of the financial and the environmental costs of conducting any single department's business.

Often environmental initiatives will save the institution money overall, but the savings accrue to one department while another department must pay a cost or expend labor to implement a program. For example, the efforts of Tufts' dining services staff to separate corrugated cardboard for recycling and purchase a baler to consolidate the material resulted in cost savings for the buildings and grounds department, which saved on waste disposal, yet the daily labor costs were borne by the dining department. Without an understanding of the full costs of environmental stewardship efforts, there is often little incentive for any one department to undertake the expense or the effort. Universities should find ways to encourage and reward waste-cutting initiatives by considering them on their full cost and benefit to the institution rather than to individual departments or schools.

**The Shortage of Capital** As in any other institution, only a finite amount of capital is available for environmental policy and technology improvements, even if the projects have rapid returns. Furthermore, emergency or higher-priority projects can sometimes tie up the available cash. Universities are generally reluctant to issue bonds or borrow money for capital improvements that don't involve major renovations. Although university endowments may have massive reserves from which an institution can borrow, few universities allow borrowing from the endowment for environmentally related projects, probably because these investments are not well understood and are perceived as unreliable as compared to stocks and bonds.

**Return Savings to Other Environmental Projects** When environmental efforts result in savings, returning a portion of the savings to fund other environmental initiatives helps keep the initiatives self-funded. While this concept is tenable, implementing it can be difficult since savings, mostly in avoided costs or avoided liabilities are difficult to quantify. Public institutions may be prohibited by law from repaying capital expenditures from operational (monthly utility or disposal charges) accounts because utility costs and capital expenditures are appropriated separately in the state legislatures. However, in private schools, experience shows that savings from environmental projects are difficult to quantify and are rarely used to seed new projects. As a result the savings return to central budgets. Where it is allowed, universities can explore paying back capital costs from the savings that accrue to operational budgets.

**Financing Alternatives** Shared savings plans, leases, and grants are among the financing options that may help universities carry out environmental programs. With a shared savings plan, a third party finances the upgrade project, and the university pays back the loan based on the calculated savings that result from the efficiency improvements. Shared savings programs are used extensively in electrical efficiency programs. Energy saving, alternatively fueled, or solid waste disposal equipment can be leased rather than bought, thereby reducing the capital needed up front. Gifts and grant money from local and national governments, private companies, and foundations are sometimes available to cover equipment or initial investments, although grants are typically small and earmarked for equipment. The University of Illinois was able to secure over $600,000 in funding from the state to cover the costs of its recycling program for five years. On a much smaller scale, Indiana University at Bloomington received $6,000 to start its recycling program.

**Finding Connections to Other University Priorities**

At corporations throughout the country, institutional commitment to the principles of pollution prevention has widespread benefit: public relations, cost reduction regulatory compliance, and worker productivity. Although a university's environmental impacts differ from those of a manufacturing company, the experience of the for-profit sector should serve as an example for university action.

Environmental concerns can complement existing university priorities. Likewise, environmental stewardship can, in part, be included in existing university structures, such as committees on planning, safety, renovations, and parking. Tufts CLEAN! staff took advantage of open invitations to the general university community and joined the traffic committee and the university facilities planning committee, among others. In the process, we learned a great deal about university priorities and were able to make suggestions and proposals that fit within existing structures.

**Cost Containment** Containing costs is a priority for most universities, so environmental leaders should point out where their efforts complement this goal. Identifying aggressive waste reduction opportunities is usually the best short-term strategy for containing costs and meeting environmental objectives.

**Competitive Advantage** Few universities have taken full advantage of the opportunity to attract students and funding through environmental stewardship action, perhaps because no university is yet able to declare that it is a truly green university. Nonetheless, current and potential students, alumni, and foundations are likely to be attracted by innovative and progressive programs. In seizing the competitive advantage, it is important that promoters of environmental achievements are accurate and humble, consistent, and open to new ideas. Part of what is required in the environmental stewardship movement is openness to ideas and a constant willingness to reexamine decisions.

**Community Building** Most colleges and universities benefit from the loyalty and pride of students, parents, alumni, and employees. Environmental stewardship efforts, especially those that are campus-wide, can help build a sense of community and purpose. Environmental leaders can use these positive aspects to their benefit through publicity that recognizes important and innovative efforts and provides positive messages to the university. For example, Tufts CLEAN! used the university's monthly staff newsletter to publicize issues and actions that were underway or completed.

**Time** Time is a limited resource for nearly everyone, and university faculty and staff are no exception. Many university projects (not just environmental projects) suffer from time constraints.

Students are passionate about environmental issues, and this passion, if channeled, can save time. For example, university purchasing departments can benefit from student commitment to environmental concerns by putting interested students to work on identifying, researching, and testing products made of recycled fibers, packaged with less material, or manufactured locally. Independent study or environmental studies students are a ready resource for university managers with limited time as well.

The university's calendar can work to the advantage or disadvantage of environmental stewardship projects. The start of a term is often the freest time; exam time is the most difficult, for both students and faculty. Grounds and dining staffs work long hours before graduation and other special events. Purchasing and accounting departments do not have a spare minute as the end of the fiscal year approaches. And many students and faculty are not on campus in the summer. Planning environmental projects to complement the calendar will make them more successful.

**Pilot Programs**

Problems arise in even the best-laid plans. Pilot programs to test the effectiveness of a program, new piece of equipment, policy, or other changes on a small scale can be invaluable for identifying unforeseen problems and working them out before a program is instituted university-wide. Departments or environmental leaders can test new equipment in a single location before installing it throughout the campus. Pilot programs can test new policies or procedures and use employee and student feedback to revise and improve programs before they are implemented on a widespread basis.

**Publicity, Involvement, Information, and Rewards**

Well-placed publicity can help to motivate and reward individual action, as well as inform the university community about new policies. Rewards are helpful for motivating change. Easiest and perhaps often overlooked is a letter of thanks to the helpful person and his or her supervisor. Students respond well to t-shirts, mugs, and even green socks for recognition of volunteer service or other jobs done well.

At Tufts, we tried a number of publicity vehicles, and each was successful in raising consciousness and increasing participation in environmental stewardship activities. Early in the project, we wrote a weekly column of tips for action in the student paper and later, a monthly story for the staff newsletter to highlight actions and accomplishments by individuals at the university. We also experimented with a newsletter, available electronically and on paper, with stories that highlighted campus action and provided action suggestions. In each case, staff commented that they were energized by being featured or by seeing others featured for taking actions similar to their own. In addition, the publicity materials provided us with opportunities to evaluate and test new programs by requesting volunteers or gauging interest in other ways. We also produced a university environmental poster, intended to provide motivation and inspiration. Throughout the duration of the project, we were pleasantly surprised to find it hanging in unlikely places as a source of inspiration and personal commitment. Many campuses work with dining services department to create a reusable mug with an environmental message as well.

Publicity and rewards need not always be one way. Some corporations have tried to elicit cost-saving ideas through the use of employee contests and suggestions that offer cash rewards. E-mail and suggestion boxes are more passive but are effective if the suggestions receive prompt responses. For example, West Chester University selected projects from suggestions submitted by members of the entire university community to develop a targeted environmental action implementation plan.[13] In April 1996 Tufts ran an environmental contest and attracted 120 ideas for reducing waste, recycling, and improving energy efficiency and water conservation. Many of the ideas were valid, but few of the entries contained enough detail about the implementation of the project to warrant the cash prizes offered.

**Setting Priorities**

Setting priorities for implementation and focusing on a target area, such as energy or waste, will help environmental leaders accomplish something. Environmental committees can be helpful in setting priorities, determining the most achievable strategies, or coordinating the actions occurring simultaneously in several departments. Often priorities are set by picking what is perceived as the biggest problem, but this problem may not be solvable with the resources, expertise, and time available. To embark on a project because it is big and needy may overlook the several projects that can be realistically accomplished and the momentum created that may, in time, put the bigger projects within reach.

At Tufts, large water cannons were used to spray water across the fields, resulting in large losses to evaporation. However, the grounds manager had spent nearly twenty-five years perfecting this watering system and was uninterested in changing it. On the other hand, we discovered that the dining manager was willing to have us work with her staff to develop some water-saving techniques, such as changes in pot washing techniques. The theoretical potential to save water in the kitchen was much less than on the athletic fields, but the real potential was greater in the kitchen because the project could be accomplished.

Planning projects comprehensively helps to avoid accomplishing only the easiest part while leaving the more difficult portion undone. Projects to upgrade lighting to improve efficiency run this risk since the projects with very rapid financial payback are often done first instead of coupling them with initiatives with longer paybacks.

Projects should build on one another, and environmental projects can do this as well. However, the projects that will succeed on any given campus are unlikely to follow such a nice progression.

## Conclusions

Institutions of higher learning have varied and often large environmental impacts. The changes that are required to minimize or eliminate these impacts are complicated and require participation and commitment at all levels. Formal committees, policy statements, and individual leaders will help make change happen. Nevertheless, each member of the university has a role to play in the efficiency, waste reduction, safety improvements, and long-term or life-cycle thinking of environmental change.

In theory, change is easy; in practice, it takes years to effect. Careful documentation of data and rigorous, realistic economic calculations are helpful tools. In addition, university environmental leaders need to remember that the process can take time and that the answers to all of our environmental choices are not known. As new understanding of these issues emerges, priorities and solutions will change. Universities are at the forefront of this evolving understanding.

# 3

# Building and Grounds

The construction, maintenance, operation, and renovation of dormitories, classrooms, offices, athletic facilities, kitchens, and their surrounding grounds hold the greatest number of meaningful opportunities for improving a university's overall environmental track record. These functions are usually the responsibility of the buildings and grounds (B&G) department (sometimes called facilities management). The ways in which the department makes repairs, renovates buildings, responds to complaints from the university community about the heating and cooling, handles trash, stores fuels and oils, and manages wastewater all have environmental consequences. B&G's influence extends to the design and construction of new buildings and the renovation of existing spaces, as well as to the maintenance of the university's trees, gardens, lawns, and athletic fields.

In most universities the B&G functions are centralized, meaning that the department has responsibilities for the physical plant and grounds that are used by many departments and university schools. This allows B&G to implement campus-wide programs or undertake efforts that affect a number of departments or schools in ways that departments on their own cannot. For example, B&G can use the savings from a lighting upgrade in the office of the psychology department to finance a lighting audit for the chemistry building.

Although many environmental actions rely on individuals' repeated choices or habits to be effective, B&G departments can make sensible technical changes that decrease the reliance on behavior changes by individuals. For example, rather than asking students to turn off the water when brushing their teeth, B&G can retrofit faucets with water-saving devices such as faucet aerators in order to save water. The difficulty for B&G, however, is that faculty, staff, and students typically are far removed from the daily problems, the technical problems, and the reality that many college and university buildings are old and have antiquated systems that are costly to run and fix. These faculty and students are also slow to see their own personal connection and responsibility to environmental action. Furthermore, the complexity of the university's facilities is often not understood, so environmental advocates may fail to recognize the nature, limitations, and possibilities for appropriate environmental actions.

The business of running a university's physical plant is complicated and rooted in systems (technology and processes) that may have evolved over many years. These systems are often there because they are familiar and they get the job done. Environmental leaders who are not members of the B&G staff (and often even those who are) need to recognize this complexity and understand the needs that the status quo meets. By promoting new technology and processes that meet those same needs and by working with B&G to solve day-to-day problems, environmental leaders can be effective catalysts for change. Certainly advocacy and pushing is needed in many cases, but often advocacy misses the realities of running university facilities and thereby is ineffective. At the same time, running university facilities holds the greatest promise for true greening of the campus.

The business of running and maintaining buildings offers unique opportunities for improvement. This chapter discusses the most common and promising actions in this area. It is important to remember that although the buildings at each institution are unique and have special problems, each of the actions detailed has been undertaken with success at a number of institutions. Some of the initiatives proposed represent a new way of approaching building operation and maintenance; others are simply the installation of new technologies. The important ingredient in the success of all such initiatives is commitment to the issues by individuals who manage and implement the B&G department's programs and the willingness by these same people to take risks that involve experimentation, expending funds, and change from the way it has always been done.

The environmental decisions faced by B&G departments involve a combination of technology, university policy, and individual behavior change. A university-wide environmental committee can be useful to B&G in informing the university community about changes that will affect classrooms, offices, athletic fields, and other spaces. The committee can be instrumental in supporting unpopular policies and can provide a forum for faculty members and students to point out opportunities for stewardship in their own facilities. It can also provide ways to link the resources of B&G with those of other departments to accomplish environmental stewardship goals.

## Solid Waste

Strategies to address solid waste, a visible and tangible symbol of our consumptive society, are often the first line of action for those seeking to reverse trends of deteriorating environmental conditions. Since trash disposal is almost invariably the responsibility of the B&G department and is paid for by the department, B&G managers have a vested interest in seeing that the university community generates as little trash as possible, that recycling systems are effective, and that wastes are disposed of properly. These actions can reduce safety problems and regulatory violations, as well as costs and labor requirements, by reducing the volume (and cost) of waste that is landfilled or incinerated, complying with state recycling mandates, and decreasing the infrastructure (hours, trucks, and equipment) needed to handle solid waste on campus. Of course, some B&G employees may see waste reduction efforts as threats to their job security or view recycling as an increased workload. For example, when Tufts' Dining Department decided to purchase a cardboard baler for cardboard recycling, it took responsibility for handling and transporting the loose cardboard as well as the operation of the baler. One result was that the number of times that the B&G staff needed to empty dumpsters at several dining halls was cut in half. At Dartmouth College and elsewhere, recycling programs have been incorporated into existing janitorial responsibilities.

B&G can play an active role in efforts to reduce solid waste material generation and encourage material reuse to reduce waste. It can also take leadership of educating the university community for waste reduction and recycling, public relations about solid waste initiatives, and policy directives that accompany the source reduction initiatives.

### Negotiate an Appropriate Waste Disposal Agreement

Most universities have a waste disposal agreement with a solid waste company to dispose of the waste off-site, in a landfill or at an incinerator. The agreement specifies such matters as who will haul the trash from the campus, how it will be measured, and how much its disposal will cost (tipping fees, usually in dollars per ton or per cubic yard of waste). The agreement may limit the materials that the university may include in the solid waste since some states ban the landfilling or incineration of certain materials.

Many institutions have saved money by restructuring their waste disposal arrangements to reflect accurately the way their campus generates waste; the result has been waste reduction and more recycling efforts, with dramatic financial savings. Items to include in the waste disposal agreement are covered containers (to avoid water-soaked trash and dumping by people from outside the university); larger roll-offs (large bins in which trash is stored or larger compactors to condense the trash in order to facilitate less frequent pickups); and tipping fees based on actual rather than estimated weight and accurate and frequent reporting of waste weight or volumes. Some universities, such as the University of Southern Maine, contract with their trash haulers for the collection of solid waste and recyclables in order to streamline the waste handling process. By combining the contracts, the solid waste hauler does not see a loss in business when recycling efforts reduce the tons of waste that are hauled away. Whenever a change to the waste disposal contract is made, resulting savings should be accounted for so they can be used to support additional source reduction or recycling efforts.

### Reduce Waste and Reuse Materials

Reducing the amount and volume of solid waste and reusing items one or more times before disposing of them as waste are important strategies for solid waste handling. Strategies to reduce waste by changing practices (double-sided copying and sharing newspaper subscriptions) and reuse (of pallets and other building materials) should be the first approach to solid waste handling. Waste reduction is even legislated in the Resource Conservation and Recovery Act (RCRA) as the first and preferred method of waste treatment. Strategies to reduce solid waste generation are described throughout this book (see chapters 4, 5, and 6 in particular). The B&G department can promote source reduction efforts, and staff should be made aware of existing efforts campus-wide and encouraged to find their waste reduction and reuse strategies. In addition, B&G can embrace source reduction throughout its own operations—in the department's offices, woodshops, garages, paint shops, and warehouses. An ethic of reduction and reuse will permeate the green university.

Recycling is tangible, participatory, and familiar, and so it is often difficult to get members of a university community to move beyond it to reduction and reuse. The key is to link specific, concrete proposals for action (such as making double-sided copies) with general waste reduction principles so that individuals understand the broad principle of reducing waste through their specific action, rather than seeing the directive for action as an isolated, and perhaps burdensome, request. Further, it appears that source reduction measures with an institutional (e.g., an administration policy to use e-mail for intracampus memos) or technological (e.g., the installation of a computer printer that can condense financial reports onto standard paper rather than ledger size) component may be easier to implement than ones relying solely on individual actions or commitments. However, components of efforts to reduce, reuse, and recycle work best in concert.

**Recycle Remaining Waste**

Recycling, the collection of material for reprocessing or remanufacturing into new products, is probably the most common institution-wide environmental strategy at colleges and universities.[1] Many universities are motivated to recycle by student and legislative pressures (e.g., bans on landfilling or incinerating paper or steel).

Paper, aluminum, glass, newspaper, and cardboard recycling are common; less frequently recycled materials are magazines, polystyrene, and building materials. White paper and aluminum cans (as well as deposit containers where applicable) usually generate revenues for the recycling program since universities can often sell these materials, although the markets for recyclables do fluctuate. The collection of other materials has financial benefits that accrue from avoided waste disposal since the materials are removed from the waste that is hauled off-site as trash.

Many recycling programs have some university staff resources, ranging from a full-time coordinator to student employees. Participation by the university community at large varies depending on the university's program. Factors that influence the participation rate and success are knowledge of the system, types of recycling containers available, and the efficacy of collection systems.[2]

The rate at which university recycling programs divert material from landfills and incinerators varies, with reported rates such as 35 percent at Connecticut College,[3] 33 percent at Tufts, 22.9 percent at the University of Virginia,[4] and 20 to 30 percent at the University of Wisconsin at Madison.[5] As early as 1988, Dartmouth concluded it had the potential to recycle 51 percent of its solid waste (69 percent if composting was included). It now recycles about 36 percent.[6] A few universities claim to reclaim and recycle more than 50 percent of their waste, although these measures are difficult to verify or compare because each university measures its recycling progress in its own way, with little attention to normalization. The success of such programs and the reported diversion rates are affected by many site-specific variables.[7] Characteristics of successful university recycling programs are described below.

**Recyclable Materials in University Waste** Paper products—white paper, mixed paper (nonwhite paper), newspaper, magazines, and corrugated cardboard—make up nearly half of a university's waste stream,[8] much as it does in the municipal solid waste in general, so paper recycling programs are the most important recycling programs on campuses. Recently paper recycling efforts have been stalled by flooded markets, so a commitment to purchase recycled paper as well is important. Contamination by food and nonrecyclable material of paper collected for recycling and the collection of paper from the many different points where it is generated are the biggest obstacles to recycling paper.

Cardboard packaging can make up a large percentage of solid waste, particularly in dining halls and bookstores. In fact, in dining halls, cardboard may be generated at a rate of a half a pound or more per person per meal. Cardboard is easily recyclable, and in many places there is a market for it, especially if it is consolidated by baling.

In dormitories and the administrative or classroom buildings, the largest portion of the trash is paper waste: white paper, colored paper, magazines, junk mail, and some cardboard and boxboard.

Glass recycling usually requires separating glass by color. Because glass cannot contain metals or other nonglass materials, collecting a product that is marketable can be difficult. Metals are generated on campus predominantly by the B&G and dining services departments in the form of used steel equipment, scrap steel, and steel or aluminum cans. Aluminum beverage cans are also a large part of a university's recyclable materials.

Plastics are generated from food containers and disposable dishes in dining departments and from packing materials in places such as the warehouse and laboratories, as well as in smaller quantities elsewhere. Most types of plastic are recyclable, but viable markets for most plastics are still developing.

Plastics are light but take up space; the cost of collecting and transporting such a low-density material may make a plastics recycling effort prohibitive since the market value of a product is generally based on its cost or price per pound. Consequently, it may be easier for the university to switch to products packaged in other materials. If plastics can be easily consolidated (at a dining hall, for example), B&G can ensure that the recycling contractor will accept them before beginning the recycling program.

Other materials, like construction and demolition debris generated from university renovations and new construction, are marketable and recyclable, and their bulk makes them worthwhile targets for removal from a university's waste stream. Universities can require that off-site contractors take construction debris with them for recycling. B&G should be careful that these same contractors do not bring material to campus from other, often smaller, sites for disposal. Some materials, like fluorescent lamps and automobile and truck tires, are regulated wastes in some states and should be removed from the waste and recycled in all cases.

**Recycling Program Staffing** Most university recycling programs require oversight by paid staff. Some universities hire recycling coordinators who manage the collection of the recyclable materials from around campus, conduct the education to inform the university about how to recycle, and manage the recycling contract. This coordinator evaluates markets and launches new programs when feasible and is often removed from coordination of solid waste disposal. In contrast, Harvard University has a rubbish coordinator who integrates solid waste and recycling collection and contracting. The advantage of this staffing arrangement is that it provides a holistic approach to the problem of solid waste and recognizes that recycling is part of the waste disposal solution. Other universities rely on student volunteers to run the program, although these programs are prone to break down during exams, vacations, and when key student leaders graduate. The assistant director of B&G at Dartmouth oversees their programs and uses existing custodians complemented by the college's coed fraternity system (CFS) to provide labor as part of the CFS community service requirement.

**Collecting Material for Recycling** The materials the university recycles are generated at many points throughout the campus. Thus, collection must be made easy for the students, faculty, and staff with accessible and clearly marked bins, explicit instructions, and trash containers nearby to prevent using recycling bins as trash bins. Without these three elements, material will be lost, and contamination of the collected material by material that is not recyclable (e.g., food, some plastics, or mixing of recyclables) will result. White and mixed paper collection bins must be beside the trash bins, and hampers to collect steel cans must stand in the university's kitchens. Selecting sufficient, appropriate, and well-labeled recycling containers from among the many on the market can make a program succeed.

Information from other programs can help B&G and recycling coordinators to determine which collection systems will work on a particular campus. Increasingly, university recycling programs are providing each student room, each office, and each classroom with both a recycling and a trash receptacle. Trash rooms are being revamped to accommodate recycling containers since the volume of recyclables is rising rapidly as total garbage volumes go down. At some universities, students must carry recyclables to bins in central locations. In residence halls, most programs collect bulky recyclables (cardboard) centrally and provide space there to collect donations and reusables like clothing, used stereo equipment, and other odds and ends.

The material must be collected from these generation points and consolidated on the campus by a contractor or B&G staff member who collects materials from numerous sites on campus. Often this next level of recycling collection can use existing resources. For example, early in its program, Dartmouth College demonstrated that its recycling program saved the college about $50,000 in avoided tipping fees. In 1994, the savings and avoided tipping fees from the disposal of trash, plus the sale of recycled materials, earned over $75,000 for the college.[9] In order to achieve these savings, the collection of recyclables was added to the duties of the college's custodial staff, a decision justified by the fact that the collection of recyclables was offset by decreased trash collection. Tufts has also capitalized on existing resources by using trucks returning empty to the warehouse to haul bales of corrugated cardboard back to the warehouse for storage. Many schools use student volunteers, students required by the university disciplinary system to donate hours of community service, and faculty and staff volunteers to carry recycling bins to central locations or participate in larger centralization of materials. Increasingly, however, they are hiring recycling coordinators or including recycling in the job description of an existing employee's position.

Collecting, storing, and handling deposit beverage containers (in states with bottle bills) is labor and space intensive. Most states require that the bottles be intact, with labels still affixed, so consolidation by crushing makes these materials lose their value. In addition, separating deposit from nondeposit containers can be troublesome and space intensive. Some contractors accept mixed containers; they pay a reduced rate for the material (four cents rather than five cents per bottle) and estimate quantities based on the size of the collection bin. In other locations, charity organizations or homeless self-help groups collect and transport the deposit containers in exchange for the proceeds.

Recycling programs must collect material in a condition that is acceptable to the market. Sometimes beverage and other food containers must be rinsed or even washed, and labels must be removed. Rinsing food containers has the added benefit of reducing the extent to which vermin are attracted to the recycling area and may need to be a standard part of a recycling program.

Materials may be collected centrally for donation to local charities. Central drop-off bins are common ways of collecting material throughout the year. End-of-semester clean-outs are also good ways to gather a great deal of material that otherwise might be thrown out.

**Consolidating Recyclable Materials** Consolidating recyclable materials by baling or crushing reduces storage space, improves the efficiency of handling recyclables, and improves the marketability of the recyclables collected. Joining forces with other departments or other nearby institutions has helped many universities pool materials and storage capabilities to make programs work. The cost of baling and crushing equipment and appropriate storage and transportation equipment can be an obstacle to implementing a working recycling program; however, the volumes of waste material generated across the university may be large enough to justify purchasing equipment for compacting, baling, or otherwise consolidating material.

When the volume of cardboard is significant, there must be a way to store and handle it efficiently. For example, a baling machine can compress and package the cardboard for easier handling and transport. Depending on the type of baler, the bales of cardboard weigh 300 to 1,000 pounds. Larger bales are harder to handle but can be more marketable because they are ready for the mill where the material will be reprocessed and do not require additional handling. Balers, which cost from $2,000 to $10,000, take up space on the loading dock, so it is prudent to locate it centrally, with easy access by all cardboard generators. Tufts University's baler stands on the loading dock of the largest dining hall. Dining supply trucks haul the cardboard from other halls to this central location, and B&G drivers pick up cardboard from other departments and haul it to the central location. Purdue has a mobile baler for cardboard that is mounted on a campus truck, and Clark University has a dedicated compactor for consolidating cardboard. Balers may also help to consolidate other materials, such as metals or plastics.

**Storage of Recyclables** A university that consolidates recyclable material needs a place to store the bales. And loose material may have to be stored as well until enough volume has accumulated to lower costs charged by the recycling company or increase revenues. More frequent pickups raise the cost of recycling. For example, a full trailer truck contains between thirty and forty bales, so storing this amount of material will reduce the frequency with which the recycler must pick up the material.

**Safety** A number of safety issues should be considered in developing any recycling program. Consolidation of material by baling or crushing results in concentrated—and heavy—hampers or dumpsters of recyclable materials. Small containers are preferable to large, heavy ones because they can be handled safely by hand. Large containers or bales are impossible to handle without the aid of a hand truck or forklift.

Workers need proper safety awareness and training to reduce injury from opening cans, breaking glass, and operating baling and crushing machinery. Crushing cans by hand or by stepping on them is unsafe. Also, washing cans and bottles to reduce the attractiveness to rodents can risk cuts and scrapes.

**Awareness and Education** Recycling programs are always changing as new materials are added to the program or new restrictions are written. Thus, recycling programs need standard ways to communicate changes to the campus community as well as to reinforce the details of ongoing programs. Regular stories in the campus paper, flyers on tables in the dining halls, and news spots in the staff publications can be helpful. Students are ready resources to assist recycling coordinators in getting the word out and can explain the recycling program to students in individual residence halls and offices throughout a campus.

Working with staff to implement environmental initiatives such as recycling may require that special and ongoing training be provided. At some universities, language may be a barrier to full or active participation in recycling programs. Harvard University and the University of Colorado translate their recycling instructions into several languages for university staff.

**Compost**

Composting organic matter—leaves, chipped brush, grass clippings, garden trimmings, and even food and animal waste—is an effective way to manage wastes either on- or off-site. Composting is a biological process that breaks down organic matter into valuable material for improving the soil quality in gardens and fields. It can help avoid negative environmental impacts associated with landfilling and incineration. Return of the compost to university grounds can improve the soil structure and its ability to retain water and reduce the need for fertilizers.

Many university B&G departments compost their organic matter. An on-site composting program requires a site that has good drainage and is somewhat isolated from neighbors or others who may be concerned about the occasional odors. If the university plans to compost its own waste on site, a front-end loader or other heavy equipment is needed to handle the compost. If an off-site contractor is used, collection and storage of material on-site and transportation of the material to the off-site location need to be provided.

Tufts found that off-site composting was a more effective way to deal with yard waste because of the campus's proximity to residential areas. The University of Maine at Orono has a composting program that captures the methane gas generated as the compost decomposes and converts it to electricity, as many landfills now do. Dartmouth University collects kraft-type paper towels used in normal handwashing in reused kraft paper shopping bags from selected administrative bathrooms and adds these towels to the compost as a bulking agent. These paper materials contain cellulose that is already processed and works exceptionally well to hold moisture and aid the composting process.

Food waste composting is more labor intensive and is regulated in many states. Ithaca College has a comprehensive program to compost food waste—about 5 tons per week—by mixing it with wood chips. The compost piles are housed in a large corrugated-steel building for five weeks, before curing out of doors for nearly a year.[10]

## Energy Management

The largest adverse environmental impacts created by universities are probably the air pollutants and carbon dioxide emissions generated by burning oil and natural gas for heating and cooling buildings and heating water and the impacts of electricity generated by a local utility on behalf of the university. The mining of oil and gas and their transport and combustion also have serious implications off-site for air quality, water pollution, climate change, and ozone depletion. Electricity in university facilities runs equipment such as heat pumps, motors, air-conditioners, lights, and laboratory equipment. Administration office buildings, athletic departments, classrooms, and laboratories are generally the largest users of electricity on a per-square-foot basis.

There are many opportunities for B&G to reduce energy use. Some of these opportunities require sophisticated equipment and training, others require diligence and frequent inspections, and still others can be accomplished by installing new technology.

Regardless of whether the source of heating and cooling is a university-wide steam system generated by a central boiler, steam purchased from a large municipal utility, or furnaces within a building, many colleges and universities have overheated buildings or inefficient heating systems. Poor heating not only is inefficient but also provides a disincentive for individual conservation efforts. Repeatedly students and employees grumble, "Why should I recycle or bike to campus if I arrive to a room that is steaming hot?!" During a recent university-wide environmental contest at Tufts, the energy manager received dozens of telephone calls pinpointing heating complaints throughout university buildings. New technology, temperature controls, and building monitoring systems offer many opportunities to decrease heating and cooling bills through improved efficiency, although the magnitude of this problem on many campuses means that visible progress is often slow.

Major improvements to the heating and cooling systems in a university are costly, especially in older facilities, built when energy was inexpensive, but the savings can be dramatic. These efforts also save natural resources, reduce operating costs, and improve building occupant satisfaction. In some cases, improving the efficiency of heating systems may reduce the amount of fuel burned sufficiently to reduce the university's reportable emissions under the Clean Air Act of 1990, thereby reducing paperwork and permitting requirements (as well as the actual air pollutants).

The actions to realize the maximum energy efficiency savings require a combination of technology, policy, and personal stewardship. The maintenance or renovation of building systems has these goals:

- Provide a comfortable living and work environment for building occupants.
- Provide services in a cost-effective manner.
- Operate systems as efficiently and cost-effectively as possible.
- Eliminate coolant leaks and minimizing air pollutant emissions.

### Hire an Energy Manager

Almost any university can benefit by creating a staff position that focuses on energy management. The energy manager oversees the most efficient operation of existing systems and has particular responsibility for improving the efficiency of the campus energy systems and policies. This person oversees renovations and new construction to ensure that they are as efficient as possible. In addition, he or she manages contractors, consultants, and financial arrangements that address energy concerns.

There are many opportunities throughout the university for reducing energy use that have high financial payback, but in the absence of dedicated attention, these actions are overlooked or are a low priority. Small colleges may be able to assign the energy manager's responsibilities to a senior engineer or plant manager. The important point is to ensure that energy management issues and policies are appropriately and clearly assigned to a member of the B&G staff.

### Conduct a Commonsense and Technical Assessment

The heating, ventilation, and air-conditioning (HVAC) systems and lighting systems in the buildings in the institution should be assessed to identify the most important and significant places for efficiency improvements. This assessment will help the energy manager determine what systems could be cleaned, improved, replaced, or redesigned to increase efficiency. Since building systems, particularly large or centralized ones, are complicated, it pays to have a professional assist with this work.

The assessment process provides an opportunity to question current operating procedures, building policies, and maintenance and to identify technical improvements and should take into account how the building is used. For example, an audit may focus on replacing a library's heating ducts while overlooking the fact that large bookcases block the air flow from the ducts or that the university does not regularly clean and maintain the existing ducts. A good auditor can examine the facilities thoroughly enough to determine the commonsense and technical solutions. Depending on their size, the HVAC systems can be quite complex. The sources of heat and cooling loads in a building are also complex, resulting in situations such as computer rooms that require cooling in a New England winter.

Campus audits are popular with student environmental clubs across the country. In these audits, the students collect aggregate data about their university or college's use of utilities and draw broad conclusions. Engineering students can be very helpful to B&G in spotting some commonsense problems, and mechanical engineering students can engage in class projects that may be appropriate parts of an energy assessment. With training and supervision, students can provide valuable assistance with the audit. B&G managers can channel student enthusiasm into a meaningful low-tech audit that can identify problem areas and quick-fix solutions. B&G can encourage students to:

- Check actual building temperatures, thermostat locations, and settings.
- Identify doors that are regularly ajar or fail to close automatically.
- Identify entrances where air locks are not functioning properly or are needed.
- Record locations where windows are open in colder months.
- Identify storm windows that are not being used or need repair.
- Evaluate the record of complaints to determine spaces that are regularly overheated or underheated.

A sophisticated audit may use trained professionals and technologies such as infrared photography to:

- Assess the condition and leaks from heating distribution systems (such as steam pipes that run beneath campus grounds).

- Determine the efficiency of HVAC systems.
- Identify opportunities to substitute heat pumps for electric resistance heat.
- Evaluate heat exchangers to reclaim heat from vented air.
- Determine the efficacy of cogeneration, ground source heat pumps, heat recovery, or solar hot water heating systems.
- Identify where technologies such as variable air volume systems for handling air are useful.
- Improve duct layout and reduce leakage from air ducts.
- Recommend energy-efficient motors and variable-speed drive motors.
- Determine system shutoff and setbacks for unoccupied hours.[11]

**Examine the Use of Resources**

An audit should examine the billing records for the institution. Most likely, these records reflect centrally purchased utilities such as oil and electricity. Meters in buildings or building complexes may measure use in smaller areas. However, on many campuses, especially those heated with central systems, it is difficult to identify the inefficient buildings because metering is done centrally.

Monitoring changes over time, where it is possible, helps determine progress on efficiency measures and pinpoint problems such as open valves and leaks. Whenever possible, existing meters should be calibrated, and new meters should be added to increase the accuracy with which buildings can be monitored.

An analysis of oil, natural gas, electricity, and water use should include some measures for normalizing the data to compare them from year to year.[12] For example, an evaluation of a dormitory's heating demand might be evaluated by the number of building occupants and the number of square feet per occupant in order to compare buildings accurately. In this way the auditor can identify the most efficient buildings and target the least efficient ones for improvement. Students in engineering departments or computer science courses are particularly adept at conducting this type of analysis.

Even minimal recordkeeping can help an auditor to:

- Account for current energy use (where and how much).
- Identify the areas with the greatest savings potential.

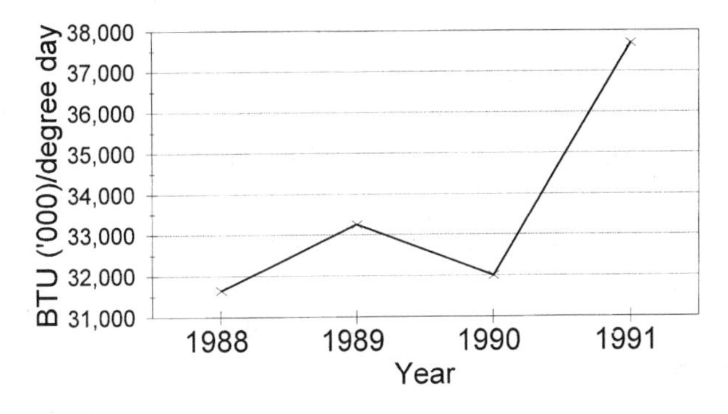

**Figure 3.1**
Heating demand at Tufts University, Medford campus

- Justify capital expenditure decisions.
- See the results of conservation efforts.
- Gain management support for energy-efficiency measures.
- Detect increased energy consumption.
- Track fluctuations in energy prices and identify the best times to purchase bulk fuel, such as fuel oil.
- Compare the energy efficiency of one building to other similar buildings.

Figure 3.1 shows the heating demand at Tufts University juxtaposed with heating degree days (a measure of the need for heat). It shows that the aggregate heat generated, as measured by fuel consumption, is not consistent with average monthly temperatures reflected by heating degree days. The fact that this university heating demand does not follow trends in external temperature indicates that there are inefficiencies in the way the building is heated, since one would expect to see higher use of heat when temperatures are colder (greater heating degree days). Information such as this can help an energy manager pinpoint areas that are most inefficient and therefore may be the highest priorities for time and money.

## Establish a Heating and Cooling Policy

A heating and cooling policy sets guidelines for building comfort and operation, provides building occupants with a set of known conditions so they can plan appropriately, and provides building operators with goals for building operation. In order for an energy policy to be successful, B&G needs to be able to maintain the temperatures described in it or to make provisions for the fact that temperatures may vary throughout a building. For example, when an office is cooler than the target temperature, B&G might allow an approved portable heating unit. Otherwise the policy will not be followed or taken seriously. The policy must be enforceable so that it can be effective. The policy also requires administrative support when the policy calls for building shutdowns or shorter hours in some buildings. Finally, the building users, especially in buildings where occupants control the thermostat, must understand the policy and follow it. The policy thus clearly sets guidelines, limits, and expectations.

An energy policy may be established by a university energy committee or another committee or by the energy manager, with approval by the director of B&G and the university operations managers. Faculty and administrative staff concerns should be considered in any policy development. No matter who is consulted, however, the policy is likely to have detractors, who want warmer—or cooler—average temperatures, regardless of the process for its development. Effective and consistent implementation, along with reasonable mechanisms for handling special situations, should be sufficient to make the policy effective.

The heating and cooling policy can include these matters:

- *Maximum winter and minimum summer temperatures.* Far too often, university buildings are too hot or too cold. B&G should strive to set and maintain reasonable and efficient temperatures. The University of Buffalo's Air Conditioning Policy stipulates a cooling temperature of 76°F during the normal work day, and their heating policy targets 68°F, although temperatures of 70°F are also common.
- *Night, weekend, and vacation setbacks.* Setting back temperatures during times when buildings receive little use can save a great deal of money. Setbacks may require coordination with the registrar in order to consolidate night, weekend, or vacation classes and activities. This was done routinely during the oil crisis of the 1970s and should be reinstituted to save money and improve efficiency. This aspect of an energy policy may be resisted by faculty who want to hold classes near their offices or laboratories or desire comfortable working conditions at night or on weekends. A university environmental committee can be instrumental for supporting B&G efforts to implement this type of policy.
- *Ban electric space heaters or air-conditioning units.* On average, 1,000 BTUs generated by electricity are three times more expensive than 1,000

BTUs generated by natural gas.[13] Because the institution pays for heat and electricity, it is reasonable to require that special heating units, such as space heaters, be preapproved by the B&G director, a process that will help B&G identify and fix problems in the most cost-effective manner possible. Moreover, space heaters can be fire hazards. Dartmouth bans them from campus.

• *Mechanisms for dealing with special cases.* Some individuals may claim to have specific health conditions, experimental conditions, or work preferences that require warmer or cooler working and living conditions. In establishing a policy, B&G must determine how to treat special cases.

The keys to a successful heating and cooling policy are the ability to meet the goals of the policy— that is, keep buildings within the temperature ranges specified by the policy—and to have some enforcement mechanism. On any campus, there are sure to be buildings where the reality of carrying out the university energy policy is nearly impossible—for example, spaces where thermostats are near doorways, so are affected regularly by outside temperatures, or buildings where a single thermostat controls many square feet on multiple floors of a building. In these areas, the policy may need to be suspended or amended to reflect the limitations of the particular building's existing technologies. B&G should acknowledge these types of problems and make a plan for improving these areas.

**Improve Heating Control**

In many older buildings, one or two thermostats control the distribution of heat for the entire building, even one with several stories and thousands of square feet of space. Since all buildings have doors and windows (heat sinks) and face both north and south, the heat needs are uneven. In addition, the load or the activity in large buildings increases the generation of heat and the need for cooling. In some buildings, thermostats may not control areas near their location or may not be functioning at all. Improving thermostatic control of the heated spaces by installing more thermostats and corresponding heating zones may save money and improve comfort by providing consistent temperature levels throughout the day and across the areas of the building.

To achieve temperature uniformity and efficiency, many universities are using remote computer energy management systems that allow a central operator to control and monitor temperatures in a collection of buildings. There is evidence that energy management systems that provide central and computerized control of space conditioning equipment can reduce energy use by 10 to 20 percent.[14] Most universities will need to contract with an outside vendor to install such a system. Where energy management systems exist, B&G managers or the energy manager should ensure that they are used effectively, that building systems do not become disconnected, and that staff members are properly trained in their use.

Sometimes central or computerized control is not feasible because installing the systems is prohibitively expensive. There is evidence that providing the building users with feedback on their energy use and education about how to reduce the use can be effective. These programs train building residents to keep temperatures low, especially at night and on vacation. Inexpensive household-type thermostats can also be installed in small buildings or houses renovated as dormitories or offices.

Thermal comfort is often linked to complaints about "poor air quality." In addition, temperature and humidity, or air condition, are often factors that contribute to indoor air contaminant levels.[15] Providing appropriate heating, cooling, and ventilation can help to improve indoor air quality.

**Upgrade Inefficient Equipment**

Changes in technology have improved the energy efficiency of heating systems, pumps, motors, distribution systems, water heaters, and the like. For example, residential gas furnaces were 63 percent efficient in the 1970s but are now as much as 97 percent efficient due to better burning techniques and insulation and improved control.[16] Water heaters are also more efficient due to better insulation, smaller pilot lights, and improved transfer of the heat to the water within the tank. And it is now possible to get twice the cooling for the same amount of energy from an air-conditioning unit.

Whenever a college buys energy-using equipment, the most cost-effective choice is usually the most efficient equipment available. Sometimes it is cost-effective and environmentally prudent to replace equipment with its more efficient counterpart even if the old equipment is properly functioning. As a rule, the older, larger, or more extensively used the equipment, the more cost-effective is its replacement.

Equipment suppliers can help a B&G manager determine the efficiency standards of new equipment. In addition, the Environmental Protection Agency's Energy Star Buildings program can provide information about relevant equipment.

**Motors** Energy-efficient motors and adjustable-speed drives can save institutions vast amounts of money by reducing electrical demand and improving the performance of the equipment run by the motor. Building systems to control heat, ventilation, and cooling are motor-controlled processes, and in countless institutions, these motors are old and inefficient; when they finally fail, they are replaced by a new motor of the same model.

Establishing a policy that requires replacement of malfunctioning motors with newer, more efficient models rather than the standard one-for-one motor replacement is prudent. Furthermore, replacing operating motors with more efficient models can often offer paybacks of under one or two years depending on the application. The U.S. Office of Technology Assessment reports that replacing existing motors with energy-efficient models can reduce motor energy use by about 3 to 8 percent.[17]

**Plug Loads** Building systems such as lighting and motors are not the whole electricity story. In buildings, office equipment is a major contributor to electricity use and heat generation. Figure 3.2 shows how new equipment in a four-building complex at Tufts affected electricity demand. The burden on a building's overall electricity use from plug loads—electrical equipment such as computers, copiers, and laboratory equipment—varies from space to space and can account for several watts per square foot. The University of Chicago found that it could reduce plug loads to 1 watt per square foot in a new building by paying careful attention to new equipment and building design.[18] (Chapters 4 and 6 describe technologies to reduce the electricity demand from some of the electrical equipment commonly found throughout universities.)

One particularly demanding electricity use results from halogen torchère floor lamps. These lamps are inexpensive and popular with students for lighting their dormitory rooms. Some of these lamps use 300 watts or more—hot enough to fry an egg in several minutes. These lamps have

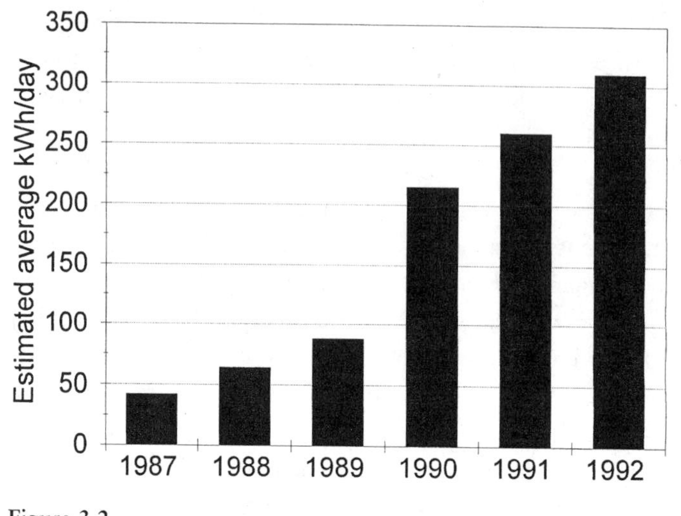


**Figure 3.2**
Cumulative daily electricity use from new equipment in a four-building complex

been blamed for fires and even death. Some universities such as Stanford have banned halogen torchère lamps, while others, such as Brown, have combined their ban with a program to supply student residences with compact fluorescent torchères. While safety is the most compelling motivation, the electricity efficiency gains are also significant.

Energy-using equipment may also contribute heavily to the generation of heat and the need for cooling. Computers, refrigerators, laboratory equipment, and ovens all produce waste heat that can dramatically affect a room's temperature. An energy management program should carefully evaluate the effect of this waste heat on the building's energy demands. Brown University found that students were adept at finding, cataloging, and finding alternatives to plug loads.

**Hot Water** Water heaters usually need to heat hot water to only about 120°F in order to meet most washing and bathing needs. Hotter water is usually diluted with cold to lower its temperature and reduce the risk of scalding. If insufficient hot water is a recurring problem, check to see that water tanks are well insulated and consider larger hot water tanks rather than raising the water temperature.

**Finance Energy Projects Creatively**

Energy projects often require capital expenditures, and there are a number of ways to finance them. Some B&G departments can take money needed to upgrade and improve energy systems directly out of capital budgets. Others set aside money saved on previous projects to pay for more efficiency upgrades. Still others are using performance contracts with energy service companies (ESCos). The ESCo puts up the cash for the equipment and the installation; the university pays back the loan from the calculated energy savings according to a contract. Dartmouth College allows borrowing from the college's working capital for energy conservation projects that pay back faster or at a rate comparable to a market rates. Loans are made from working capital, and a payment plan is established. Payments are made from operation budgets until the loan is repaid.

B&G directors may be able to borrow from reserves or capital budgets and pay them back from the savings to operating budgets. Significant long-term gains can accrue to the university and the environment if money saved from energy efficiency finances other energy-efficiency investments. For example, if thermostat controls are bought using a capital improvement budget and result in a 5 percent drop in oil used for heating, the heating budget might reimburse the capital budget for the investment. In this way, the savings can be reinvested in other projects. Regardless of the mechanism, B&G managers need to find ways to document and account for savings and return these savings to related projects when possible.

## Lighting

Lighting is estimated to use 20 to 25 percent of the electricity in the United States. In commercial buildings, about 41 percent of the electricity and 28 percent of the total energy consumed is for lighting.[19] New lighting technologies are available to provide the same or better lighting quality with fixtures that consume less electricity than those available even five years ago. One example is the improved performance of compact fluorescent lamps. With available technologies, electricity used for lighting can be reduced from 39 to 83 percent of its prior demand.[20] Even spaces

**Box 3.1**
Summary of common lighting terms

*Lighting fixture* refers to the entire assembly of lamp, ballast, filter, reflector or lens, and other parts. Incorporating reflectors and lens can effectively direct light on work spaces and decrease the number of lamps needed to light a space.
*Incandescent lamps* work on the principle of electrical resistance. Electric current flows through a wire filament, which gets hot and glows. Incandescent lamps create both heat and light.
*Compact fluorescent lamps* are usually a screw-in version of a fluorescent lamp and can replace most standard incandescent lamps. Most compact fluorescent lamps have a built-in ballast. A 13 watt compact fluorescent lamp produces the same amount of light as a 60 watt incandescent lamp.
*Fluorescent lamps* contain gas instead of wire filaments. Electrical current makes the gas atoms glow, creating light with very little heat. There are many types of fluorescent lamps. The most common are the long tubelike lamps. Of these, T-12 lamps are larger in diameter and are less efficient than T-8 lamps.
*Ballasts* are devices that charge the electrical current in fluorescent lights. Fluorescent lamps always need ballasts. Magnetic ballasts are generally less efficient than electronic ballasts.
*Sensors* detect changes in the room and trigger lighting changes accordingly. Occupancy sensors that detect the presence of movement or heat in the room and turn the lights on (or off) and daylight sensors that vary light levels according to the ambient day light available are two common types of sensors.

that use fluorescent lamps probably can use up to 50 percent less electricity with new technologies. In addition, the local electrical utility company may provide assistance and even incentives to improve lighting efficiency. Box 3.1 describes some of the most common lighting terms.

Most campuses have many opportunities to upgrade lighting to improve efficiency, and indeed, many universities have taken advantage of these opportunities. The University of Missouri at Columbia has implemented state-of-the-art energy-efficient lighting technology and is saving more than 4.5 million kilowatt-hours, and $320,000, annually with only 35 percent of its floorspace upgraded.[21] The City University of New York has estimated its savings at $3.6 million annually.[22]

Lighting upgrade projects are not without pitfalls and problems.

Among them are the need to ensure that the quality of light in offices, libraries, laboratories, and dormitories is preserved or improved (this does not necessarily mean more or even the same level of light) and that regulations such as building codes and laws governing the disposal of old lighting equipment are met. Lighting upgrade projects typically require the investment of capital, so it is important to document the savings of a lighting project in avoided costs so they can be applied to subsequent projects.

In order to be effective and achieve maximum results, the realistic and achievable goals of a lighting retrofit program include the following:

- Reduce the electricity used to light a building or a space.
- Reduce long-term operating costs for lighting, including electricity.
- Reduce labor for replacing lamps on a regular basis (relamping).
- Maintain or improve lighting quality.
- Dispose of wastes properly.

There are more difficult goals as well:

- Improve the knowledge about and records of the lighting and building systems on campus.
- Apply dollars generated from avoided costs to new retrofit projects.
- Educate building occupants about their electricity use.

Besides improving lighting quality, improving lighting efficiency, and saving money, there are a number of other often overlooked benefits of lighting projects. Box 3.2 summarizes the benefits of a lighting upgrade project. One benefit is a reduction in the cooling load—the amount of waste heat removed by air-conditioning. Since energy from lights is transferred into heat as well as light (the heat is lost energy), inefficient lighting results in a great deal of waste heat. In warm weather, and especially in buildings with many heat sources (such as computers and other office equipment), waste heat increases the energy required to cool the building, so reducing inefficiencies in lighting can reduce air-conditioning needs. In cool weather, lighting helps warm the building, reducing the amount of heat needed for a heating system, but it is almost always cheaper and more environmentally responsible to get heat from a central boiler intended for that purpose than to get it from light bulbs.

**Box 3.2**
Benefits of lighting upgrades

More efficient lamps
More efficient light
Lamps with longer life
Fuller-spectrum light
Cleaned fixtures
A lighting inventory
Removal of PCBs
No flicker or hum
Reduced heat load
Reduced demand on transformers
Increased available transformer capacity

A second benefit of upgrading the lighting is that when electricity demand in a building is reduced, the transformer that services that building effectively gains available capacity—that is, the ability to serve more buildings or spaces. Therefore, lighting upgrades can be a cost-effective way to gain transformer capacity that may be needed for facility expansion or new equipment and may often be more cost-effective than purchasing a new transformer alone.

A third benefit is the increased environmental awareness of the building occupants. If coupled with an outreach effort, lighting upgrade programs can be public relations boons that bring awareness of environmental issues to the fore. The Park Plaza Hotel in Boston implemented a range of environmental improvement efforts including the use of compact fluorescent in its rooms. Publicity about the program became a selling feature for the hotel. Universities can gain similar benefits from increased enrollments and favorable press.

## Join the EPA Green Lights Program

The Green Lights program is a voluntary program to promote energy efficiency through lighting retrofits. The university signs a memorandum of understanding with the EPA committing the signatory to upgrading its lighting and improving its efficiency in 90 percent of its built space up to the point that it is "profitable." For its part, EPA provides basic technical assistance as well as access to computer software, regular newsletters, workshops, and consultation. More than 120 educational institutions are signataries to the Green Lights agreement, including Tufts, Columbia University, Bucknell University, Fisk University, and Yosemite Community College.

Green Lights partners (those signing the memorandum of understanding) make a public commitment to energy-efficient lighting and undertake a program to make it happen. Signing the Green Lights agreement is an opportunity for publicizing the institution's commitment to energy efficiency and its associated cost avoidance and to the environment. In order to maximize the power of this agreement, a top university administrator should sign the agreement. The energy manager or other environmental leader can use the process of signing the agreement to help administrators understand the benefits (and capital investment) of this commitment. At Tufts the commitment to Green Lights was a catalyst for action and an important demonstration of commitment.

**Conduct a Detailed Audit**

Upgrading the lighting will be most successful when it is planned carefully and carried out in the same manner as other capital improvement projects. A lighting program encompasses all the lighting factors, such as light level, types of lamps, indoor and outdoor lighting, and lighting control.

Those who are conducting a lighting upgrade should be wary of consulting with building occupants. Although the lighting plan needs to consider the needs of the place, this can be done by determining the uses of space and asking questions only of those people who manage those activities. A general query to all users about lighting needs is certain to elicit responses from many who believe that they *must* have incandescent lighting and that the existing fluorescent lighting hurts their eyes. It is a fallacy that eyestrain results from fluorescent lighting. (Their space may have insufficient or poor distribution of light, however.)

There is a temptation in installing efficient lamps to replace the existing lamps one for one. Certainly some savings will result by simply replacing a lamp with a more efficient model, but these savings can be dramatically increased by carefully considering the entire light fixture, its placement, and the reflectors and lenses that accompany the lamp itself. The steps of auditing and thinking through a comprehensive program are the secret

**Box 3.3**
Elements of a lighting audit for existing rooms

Existing conditions
- Light levels (in foot-candles)
- Electricity demand (in kilowatts)
- Electricity use (in kilowatt-hours)

Lighting needs
- Necessary light levels in foot-candles for existing or planned room use (should specify height above the floor)
- Special needs, such as stage lighting

Lighting design
- Proposed new lighting (lamps, ballast, reflectors, and lens) and resulting light levels
- Proposed lighting control (occupancy sensors, daylight sensors, or energy management systems)
- Proposed changes in wall paint color (white versus black)

Costs and savings
- Resulting electricity demand, use, and cost
- Labor cost
- Disposal costs (ballasts and lamps)
- Available rebate from the utility
- Summary of costs and savings from avoided electricity use (including assumptions about hours of use, hourly rates, and kilowatt-hour costs)

to long-term rewards, improved lighting quality, and greatest efficiency. To maximize the efficiency gains and related savings, as well as improvements in lighting quality, B&G can work with an individual or firm with experience in lighting efficiency and design to conduct an audit. Box 3.3 shows the components of a lighting audit. Electricity conservation and financial savings are particularly high in applications such as exit signs, incandescent task lights, and areas where there is excessive light or glare.

Hiring an auditor for the comprehensive audit and development of the lighting plan requires an initial expenditure for a study, and administrators may be reluctant to spend money in this way. Some lighting contractors provide a free audit as a bid to install efficient lighting. This usually does not work well since contractors who wish to do the final installation work may identify only the projects with the fastest payback in order to make their proposals appear more attractive than a competitor's.

Some school systems have successfully used students to conduct some simple audit tasks. Certainly students can be valuable, particularly in locating certain types of lighting fixtures and incandescent bulbs. But students and other environmental leaders need to be carefully trained to assess lighting levels before their lighting audits can be considered complete.

**Upgrade the Lighting**

Some B&G departments can use university staff to implement the lighting plan, but most will need to contract with an outside firm to install the fixtures. The contract and work plan should encompass installation, labor, clean-up, and waste disposal. The university energy manager should oversee the installation of new lighting fixtures.

In many instances, it will be prudent to try lighting upgrades on a pilot basis in one room or one floor before proceeding to upgrade an entire building. The assumptions about building use, sunlight, and the reflective surfaces of the rooms are important contributors to the success of a well-designed lighting plan. Special considerations such as the durability of lens-covering lamps in dormitories may need to be tested before substantial investment is made in a new model. Trial installations in a single room can also be helpful in buildings where there will be resistance to change. Trial installations can demonstrate that new lamps (size T-8) with electronic ballasts do not flicker or hum, common complaints about standard fluorescent lamps (size T-12) with magnetic ballasts.[23]

**Dispose of Wastes Properly**

Regardless of who installs the new lighting, the university will be liable for the disposal of any wastes that are regulated. Magnetic ballasts manufactured before 1978 contain polychlorinated biphenyls (PCBs), which are carcinogenic and can leak from ballasts when they are damaged or crushed (as in a trash compactor). Ballasts made after 1978 are PCB-free and are labeled "no PCBs." Unless a ballast is so labeled, it should be assumed to contain PCBs. Depending on the region of the country, non-leaking ballasts are regulated differently. However, leaking ballasts, including those that were not leaking when they left the university (e.g., those crushed in the course of ordinary solid waste disposal), that contain

PCBs must be disposed of as hazardous waste (at a cost of about two to four dollars per ballast).

Failure to dispose of these wastes can involve an institution in costly cleanup costs, including liability and fines. In fact, an institution that contributes even a small amount of hazardous waste to a hazardous waste site can be responsible for all or part of the larger cleanup of that entire site. Superfund sites such as the contamination of New Bedford harbor (in Massachusetts) are testimony to the damaging effect of improper PCB disposal and the financial liability that is incurred by institutions that contribute to the site.

Fluorescent lamps too must be disposed of as hazardous waste in some states, although in most states they can legally be disposed of as trash. Lamp recycling by companies that recycle the glass and metals and reclaim the mercury from the tubes is the responsible alternative to disposal of fluorescent lamps as trash.

### Educate Building Users

Informing building occupants about the lighting upgrade is an educational opportunity. Properly done, it provides people with a sense of how they can save electricity and reduce their electric bills at home as well. Some publicity programs have made use of the conservation material available from the university's electric utility company to provide information to building occupants. Educational efforts around lighting upgrades are also a good time to remind members of the university community to turn off their lights when they are not in use.

## Indoor Air Quality

Good indoor air quality (IAQ) is an important component of the green university. IAQ includes the introduction and distribution of adequate ventilation, control of airborne contaminants, and maintenance of acceptable temperature and relative humidity.[24] Health problems resulting from contamination of a building's indoor air are being recognized increasingly as valid working condition problems. In fact, in the future, workers may be able to file workers' compensation claims for health problems related to poor IAQ. As many as 40 percent of the buildings in this country may have air quality problems ranging from minor to severe. Recently, the University of Massachusetts in Boston had to close its buildings for a short time because many staff and faculty were complaining of headaches and nausea, probably due to poor air quality resulting from inadequate ventilation or an indoor contaminant, or both.

IAQ problems in universities have many sources. Outdoor sources, such as pollen and pollution, vehicle exhaust, odors, and moisture, may enter the building. HVAC equipment can contribute to IAQ problems when there is dust, dirt, or microbiological growth in the system. Improper ventilation of combustion products can also cause problems. Other sources of poor IAQ include office equipment such as copy machines that use solvent and toners; office furniture and carpeting that release volatile organic compounds and ozone; human activities such as smoking; cleaning supplies; and poorly ventilated laboratories.

Maintaining and improving air quality is usually not simple or well defined. Rather, it is a changing and complex issue that must be examined case by case and building by building. B&G, in coordination with the environmental health and safety department, should acknowledge the legitimacy of air quality complaints and take steps to address them. The goal is to identify the sources of indoor pollutants and curb them by stopping the activity or isolating it. Preventing problems in the first place (during building construction) is particularly important and usually the most cost-effective solution. Rectifying indoor air problems often has other benefits, such as improved comfort, increased safety, and increases in staff and student productivity. Table 3.1 shows some of the major sources of IAQ problems in universities.

**Determine the Basis for IAQ Complaints**

IAQ complaints take many forms, from headaches to fatigue, nasal congestion to nausea. There are many potential sources of IAQ problems, so it is important for B&G, often in cooperation with environmental health and safety department staff, to gather detailed information about the complaints in order to pinpoint the source of the problem. Much of this information involves finding abnormal trends or conditions, so it relies on a knowledge of normal building conditions. B&G staff should therefore be involved in any evaluation of IAQ problems. The following

**Table 3.1**
Sources of indoor air quality problems in universities

| Pollutant | Sources | Potential health effects | Control measures |
|---|---|---|---|
| Combustion by-products: nitric oxide, carbon monoxide, carbon dioxide, odors, particulates | Smoking<br>Routine odors from occupants and normal activity<br>Unvented gases<br>Odors coming in the air intake vents<br>Any heating source<br>Car exhaust | Lung cancer<br>Asthma<br>Breathing disorders<br>Drowsiness<br>Headaches<br>Discomfort<br>Respiratory infections | Ban indoor smoking in all university spaces, including private offices<br>Offer smoking cessation assistance<br>Improve ventilation<br>Ensure that ventilation meets current standards<br>Ensure that outdoor air sources are not near parking lots or sources of fumes<br>Reduce occupant density<br>Relocate or reduce use of heat-generating equipment<br>Ensure that ventilation is not blocked |
| Biological contaminants: molds, fungi, bacteria, mildew, allergens | Humidifiers<br>Air-conditioners<br>Standing water on roofs near air intake<br>Dust mites, cockroaches, rodents | Allergies<br>Headaches<br>Flu symptoms | Take steps to reduce moisture<br>Maintain relative humidity between 30% and 40%<br>Decrease heat losses from exterior walls (insulate)<br>Clean air-conditioner and humidifier filters<br>Exhaust fans in bathrooms and kitchens<br>Integrated pest management |
| Asbestos | Wall and ceiling insulation installed between 1930 and 1950<br>Old pipe insulation | Skin irritation<br>Long-term inhalation can lead to lung cancer | Maintenance and in situ management<br>Avoid asbestos material<br>Remove friable asbestos using a licensed contractor |

| | | | |
|---|---|---|---|
| | Some older vinyl floor tiles<br>Old fireproof cloth products | Asbestosis (scarring of the lung tissue) | Use appropriate in-place management that contains and does not disturb the asbestos<br>Avoid exposure |
| Radon (a radioactive gas naturally occurring in all soil and rock) | Soil and rock<br>Seeps into the building from natural sources | Lung cancer | Test to determine levels<br>Seal cracks<br>Ventilate basement<br>Retest |
| Volatile organic compounds (hydrocarbons) | Cleaning products<br>Propellants for aerosol products<br>Deordorizers<br>Paints and thinners | Irritation of mucous membranes of nose<br>Headaches<br>Heartburn<br>Mental confusion | Use products according to instructions, especially in correct concentrations<br>Mix concentrated solutions centrally in a controlled manner<br>Use products with plenty of ventilation<br>Switch to nontoxic alternatives<br>Store and dispose of products properly |
| Semivolatile organics: formaldehyde (HCHO), PCBs | New carpeting (usually in the backing)<br>Furniture<br>Particle board<br>Adhesives<br>Urea formaldehyde insulation | Eye irritation<br>Upper and lower respiratory irritation<br>Pneumonia (sensitivity varies widely) | Remove source if identifiable<br>Steam clean new furniture and carpets<br>Avoid products with high levels of formaldehyde<br>Seal particle board with varnish or vinyl wallpaper<br>Ventilate |
| Chemicals | Laboratory experiments<br>Chemicals sitting in labs<br>Chemical storage rooms | Eye irritation<br>Respiratory irritation | Store chemicals properly at all times<br>Reduce chemical use at all times<br>Ensure that ventilation is adequate<br>Keep storage rooms at negative pressure |

steps may help to gather the information needed to identify IAQ problems and determine their solution:

1. Conduct a walk-through with one or more visits to the complaint area.
2. Collect information about the history of the building and the nature of the complaints. Did the building have a different previous use? Has new furniture or equipment been installed? Have building systems been changed? What is the condition of the HVAC equipment? How are indoor plants maintained (are they sprayed)? Do closets or drawers contain materials that contribute to odors (e.g., workout clothing or towels)?
3. Ask the building occupants to fill out a log detailing their complaint, the time of day it typically occurs, and the building conditions (e.g., temperature, odors, outside weather, equipment running, etc.). This log may help to identify sources that are intermittent or seasonal.
4. Measure building temperature, humidity, air flow, and carbon dioxide levels. Experienced technicians should conduct air sampling. Regional EPA or state environmental offices can help to find reputable technicians if the university does not have trained staff.
5. Identify other problems that may cause symptoms similar to those caused by poor IAQ, such as flickering or inadequate lighting or lack of natural sunlight.

An EPA manual, *Building Air Quality,*[25] and a packet of materials prepared by the EPA and others, *Indoor Air Quality, Tools for Schools, Action Kit,*[26] provide guidance on identifying and solving IAQ problems.

**Prohibit Smoking in University Buildings and Facilities**

Smoking is one of the leading causes of IAQ complaints, and increasingly universities are creating smoke-free zones or declaring all buildings smoke free. Smoke can travel throughout a building in its ventilation system, so eliminating smoking in all university buildings, including faculty and administrators' office, is a healthy and important step.

Smoking cessation assistance for smokers who wish to quit should accompany smoking bans. In addition, universities should involve the office of international students when carrying out a smoking ban. Since the incidence of student smoking is often higher among international students, there may be valid concerns that smoking bans will segregate international student smokers from other students. Special programs may need to be developed to overcome this problem such as targeted smoking cessation efforts or special attention to the pairing of international students who smoke with those from the United States.

**Ensure Proper Ventilation**

Ventilation can help control indoor contaminants in some instances by diluting the contaminants or removing a localized problem with an exhaust system. The standards determining the amount of outdoor air adequate for proper ventilation have changed over time. Buildings constructed prior to 1989, the date of the most recent American Society of Heating, Refrigeration and Air-Conditioning Engineers, Inc. (ASHRAE) standard revision, may not have adequate fresh air ventilation.[27] Studies have shown that increasing ventilation to meet the new 1989 standards (from 5 to 15 to 20 cubic feet per minute per person) does not measurably increase energy consumption.[28] Even these standards do not necessarily ensure adequate air mixing or air quality. Sensors may be needed to determine when carbon dioxide levels are dangerous or increase stuffiness and poor air quality. Ventilation systems should be operated so that they remove pollens and pollutants from incoming air.

As a campus expands, buildings may be renovated to accommodate new uses, and parking areas and roads may be moved or changed. In the course of these changes, air handling systems are added or moved, or the intake manifolds end up next to a parking lot or a loading zone. As a result, the ventilation equipment inhales carbon monoxide and other exhaust and distributes it throughout the building. The solution may be complicated and expensive once the system is in place, so prevention is the best medicine. One office at Tufts had to request that trucks shut off their engines while they made their deliveries to avoid filling the office with exhaust.

Spaces with high concentrations of volatile organic compounds or toxics should be vented directly to the outdoors. In particular, improper ventilation in laboratories or other areas that store research chemicals may be a source of IAQ problems. Appropriate source-reduction techniques, proper ventilation hoods, and the use of ventilation systems can improve this problem. B&G should be notified where chemicals are used in order to ensure that the building systems are adjusted to avoid exposure to hazardous fumes. Renovations of laboratory space or conversions of office space to laboratories should be done with care to ensure that ventilation is adequate. Venting photocopier spaces is rarely done but may reduce stuffiness and excess heat buildup.

Problems with the thermal comfort of a room or building may contribute to problems that are characterized as air quality problems, such as stuffiness or overheating. In addition, poor heating control and ventilation can contribute to or accentuate air quality problems. Ventilation need not increase heating costs substantially. New air exchange technologies can capture up to 80 percent of the heat from exhaust air.[29]

### Identify Sources of Humidity and Standing Water

Excess humidity and moisture are common sources of IAQ problems. Puddling on roofs, leaky windows and ceilings, and HVAC condensers can be sources of mold and bacterial buildup. Basements or mechanical equipment rooms can also be sources of dampness and standing water. Poor sanitation or failure to dry out carpets adequately following flooding or other spills can cause problems from molds and mildews. Floor drains covered by carpets or cabinets can contain standing water or may have drain traps that emit odors.

B&G can reduce molds and mildew by drying, cleaning, and disinfecting areas where water and molds have accumulated, fixing leaks to eliminate moisture, and keeping building humidity low by reducing sources of moisture and properly tuning the HVAC equipment. After measures are taken to address IAQ problems, follow-up evaluations should be used to ensure that the solutions remain in place and the problems have diminished.

### Test for Radon Gas

Radon is a colorless, odorless radioactive gas produced by the decay of radium and is thought to be the second leading cause (behind smoking) of lung cancer. It occurs in almost all soil and rock and can enter buildings through cracks and other openings in foundations. Radon enters buildings from the soil or through water. Because of the exposure to soils, radon tends to affect lower floors and basements of buildings, although elevator shafts may transport the gas to upper levels. Exposure to radon gas dissolved in water is a problem when water comes from bedrock sources and greatest where radon can become airborne from showers and laundries.

Radon is as great (or greater) a problem in homes as it is in larger buildings, so university efforts at detection and remediation should be sure to include small houses and offices. Detection in small buildings can be accomplished with a residential radon test kit, approved by the EPA, which is available at hardware stores and from the American Lung Association. In larger buildings, a systematic testing plan should be established using short-term and long-term testing (more than ninety days). EPA recommends that all frequently used rooms be tested. Tests should be conducted when doors and windows have been closed for at least twelve hours, during normal weather in the colder months, and with the HVAC system operating normally.

Where radon levels exceed 4 picocurie per liter, an exhaust system or soil suction system that removes gas from beneath the foundation can ventilate it from the building, although reducing radon levels below 2 pCi/L is often difficult. Pressurizing the building and sealing cracks may prevent radon from entering and can be effective. Construction plans for new buildings should consider radon-reduction opportunities such as installed subfloor ventilation systems.

Radon exposure from water results from ingestion and inhalation of the gas as it off-gases from showers, laundries, cafeterias, and other concentrated supplies. Exposure from swimming pools and similar large water sources is insignificant since radon has a half-life of only three days. Well water, especially from bedrock sources, should be tested. There are no official standards for radon in water, although in 1991 EPA proposed a standard for action in public water supplies requiring action at levels above 300 pCi/L. Aeration or filtering can help remove waterborne radon.

## Water

Historically water has been inexpensive, but increasingly, unpolluted sources are scarce and water treatment is expensive. In arid regions like parts of California and Arizona, water rights are hotly contested, and awareness of water scarcity and the need for conservation is high. Wastewater treatment is also resource intensive and expensive, requiring large treatment plants and high technology. In the greater Boston area, water and sewer rates have risen dramatically largely due to wastewater treatment costs. Often students tend to think only about water supply, which they perceive to be adequate, while ignoring the inadequate and the costs of wastewater treatment.

Conserving water has environmental and financial benefits. Water conservation strategies are simple to undertake and require changing technology, changing behavior, and proper maintenance. Conserving water reduces the amount of freshwater consumed, the quantity of water disposed of to the sewer, and the amount of energy needed for hot water heating. Hot water heating accounts for 15 percent of residential energy use and about 4 percent of the commercial energy use.[30]

**Conduct a Water Audit**

As with energy and lighting, identifying and evaluating water-using equipment is the first step toward water conservation. Water use depends on both the flow rate and amount of time used. Data about a building's total water use are less important than data about the type and condition of water-using equipment. If possible, B&G should gather information about water use habits in showers, dormitory bathrooms, kitchens, and grounds maintenance. These data are difficult to gather and interpret, but students are ready resources to gather the information and translate it into a water conservation campaign. Students can be helpful for documenting a variety of water data:

- The size of toilets (in gallons) in each campus bathroom
- The flow rate of every shower (in gallons per minute)
- The flow rate of faucets in sinks, kitchens, hoses, and outdoor faucets (gallons per minute)
- Leaks and drips on a spot-check or ongoing basis

B&G or outside consultants can also identify the existence of water cooling systems that do not recirculate their water (illegal in many states) and help establish a routine leak detection plan.

Table 3.2
Water use in rest rooms

| Device | Traditional water use | Low-flow water use |
|---|---|---|
| Toilet (tank) | 6 gallons per flush | 1.6 gallons per flush |
| Shower | 6 gallons per minute | 2 gallons per minute |
| Faucet | 6 gallons per minute | 2 gallons per minute |

**Retrofit Sanitary Systems**

Replacing conventional toilets, faucet aerators, and showerheads with low-flow models can reduce water use in university settings that receive high use. In some state, including California and Massachusetts, building codes require low-flow fixtures in all new construction and renovations. The Copley Plaza Hotel in Boston found it cost-effective to replace all 6-gallon flush toilets with new 1.6-gallon models. The hotel benefited as well from the national publicity that was generated when it arranged to have the porcelain crushed for use as reflective material on highways. The University of Northern Iowa found that replacing 6-gallon-per-minute (gpm) showerheads in dormitories with 2-gpm models paid back for itself in six weeks![31] Table 3.2 compares traditional and water-saving devices.

Devices such as automatic shutoff nozzles on faucets can reduce water use in public rest rooms and student residences. Spring-loaded valves or timers on manually operated water outlets can prevent water waste as well.

**Conduct A Water Conservation Campaign**

Because total water use depends on the demand for water from individuals in rest rooms and dining halls, a campaign to educate users about water-conserving habits is an effective and often important part of a comprehensive water conservation effort. Local water suppliers and state water offices can help plan a conservation campaign and provide posters, stickers, motivational facts, and information about local supplies. Students are helpful to water conservation efforts by encouraging peers to take shorter showers, turn off faucets, and report leaks. Nonetheless, the effectiveness of a water conservation education effort will depend in part on the perceived local water conditions. For example, people are generally far less motivated to conserve water during a rainy spring than during an arid summer, regardless of the water available in the local reservoir—a condition often influenced by the weather patterns of several years rather than months. Education must be ongoing and comprehensive in order to be effective.

### Eliminate Once-Through Systems

Most universities' air-conditioning and heating systems use large quantities of water. Closed-loop cooling and heating systems that recycle water after sending it through a cooling tower or a steam plant use far less water than once-through systems. In some locations, it is against regulations to put once-through or noncontact cooling water into the sewer. Closed-loop systems are more expensive to install, but recycling the water may improve the performance of cooling equipment by maintaining a more constant temperature. In some cases, air cooling should be substituted for water cooling to reduce water use. Steam systems should be designed to return maximum quantities of steam condensate to the boilers.

### Fix Leaks Promptly

Dripping or leaking water pipes and fixtures, running hoses, and malfunctioning toilets together can waste thousands of gallons of water each month. For example, a leaking toilet can waste 50 gallons a day, and a fast drip from a faucet can waste 200 gallons each week. B&G managers can train custodians to identify leaks and report them rapidly and establish systems to repair leaks quickly and identify worn valves or pipes to prevent leaks. Leaking coils in hot water heaters and leaking elements in swimming pool heaters are easy to overlook for long periods, so regular leak detection inspections should be established.

### Install a Swimming Pool Cover

Evaporative water losses from swimming pools can waste thousands of gallons of water annually and the associated energy depending on the size of the pool. Installing a cover over the swimming pool during periods when the pool is not used will reduce this loss. A pool cover also helps maintain the pool temperature, reduces humidity in the building, and reduces the use of mechanical equipment to remove humidity. A common problem with pool covers is convincing pool staff to use them regularly.

**Reduce Water Used in Laboratories**

Water conservation efforts should examine each laboratory to determine if water-saving opportunities exist. For example, water aspirators are used in many laboratories to create a vacuum for laboratory experiments. In some cases they are used sparingly, but in others they run nearly constantly. In order to conserve water, aspirators can be replaced with vacuum systems that do not use water. In a biochemistry laboratory at Boston University, replacing aspirators that were used more than one and one-half hours each day paid for itself in less than three years (including electricity costs).

Another example is the reverse osmosis that is often used to produce distilled water needed by laboratories, particularly for glass washing. Some reverse osmosis systems are only about 30 percent efficient and lose 70 percent of the water as waste steam.[32] New technologies can reduce the amount of waste steam and reclaim the previously wasted water through improved filter maintenance or as condensate cooling water in other equipment such as autoclaves.

**Discharge Only Water, Sewage, and Permitted Substances**

Some universities have National Pollutant, Discharge, and Elimination System (NPDES) permits under the Clean Water Act that set discharge standards that must be met. Often these permits are required to operate a treatment facility or discharge into a municipal facility. At universities, violations for the discharge of acidic or basic material (outside allowed pH standards) and violations for heavy metals, like silver from photography labs, are among the most common problems. It can be difficult for universities, with their many independent autonomous departments, to ensure compliance, but B&G, working the environmental health and safety staff, should try to ensure that only permitted materials are discharged to wastewater systems. Discharging oil or hazardous substances should be strictly forbidden.

If the university operates its own sewage treatment plant, it should be using secondary or tertiary treatment in almost all cases.[33] If only primary treatment is in place, the university should allocate monies for upgrading the system in the near term to protect the quality of the receiving water bodies. New water treatment technologies such as the use of plants to purify water may also be appropriate for university wastewater treatment.

## Maintenance

Since most universities plan to be in business for many years, maintaining their properties in the near term makes economic and environmental sense. Yet in some universities, the cost of maintenance limits the resources that are available to invest in efficiency measures that could reduce operating costs. Nationwide, the cost of deferred maintenance at college and university buildings is estimated to be $26 billion.[34] An effective building maintenance program can help to improve the environmental performance of an institution's buildings and spot problems early.

### Train Maintenance Staff

Training can help maintenance personnel and managers to optimize building and building system performance and recognize and report problems. A training program includes basic heating control, energy tracking, identification of air quality problems, and the basics of lighting.[35] In addition, maintenance personnel might be trained to include low-tech actions in their routines: turning out lights, lowering thermostat levels, closing windows, and turning off water. Maintenance personnel can also include regular maintenance inspections and reporting of problems in their duties. This training can be carried out by B&G managers and reinforced by shift supervisors. Custodial services can also be targeted to deliver preventative cleaning such as regular dust mopping of hardwood floors, regular cleaning of the areas just inside outside doorways, and scheduling of major cleaning when buildings are empty.

Custodial staff can be the point people for identifying maintenance needs before they become large problems. Universities that empower custodians to find and report problems such as leaking water faucets or pipes, running toilets, overheated or underheated rooms, open windows, and stuck heating ducts will see a payback in water and energy savings. For example, early detection of a leaking water pipe can save gallons of water, gallons of heating oil, and several hundred dollars. Early detection of over- or underheated spaces saves heat and reduces the labor involved in processing heating complaints. The key to a successful leak detection and repair program is to provide incentives for prompt reporting and ensure that repairs are carried out in a timely manner.

**Expect Student Responsibility**

College students are tough on lights, walls, furniture, and kitchen equipment. Certainly some wear and tear is standard, but damage to dormitories is often difficult to curb and it is often even more difficult to identify the guilty parties. Establishing a system of assigning financial responsibility for campus damage helps reduce damage, fund its repair, and promote respect for property. Bates College has an aggressive program of daily monitoring of all college residences and gathering places. The maintenance staff inspects individual dormitory rooms during vacations and at the end of each semester. Maintenance crews repair damage promptly, often within the same day, and the residents on the floor are billed for the repair. If the resulting bills are unclaimed by the responsible parties, they are distributed evenly among the residents and must be paid. As a result, students take more care that damage is not incurred and have an incentive to identify guilty parties. The system is labor intensive, but it ensures that facilities are well maintained and funds are available to make the repairs.

**Select Environmentally Friendly Cleaning Products**

Universities use an array of cleaning products in the everyday maintenance of buildings and facilities: all-purpose cleaners, degreasers, floor strippers, waxes, disinfectants, carpet shampoos, and detergents. Many of these products are toxic, corrosive, or irritant—characteristics of hazardous wastes. Heavy metals such as lead, cadmium, copper, mercury, nickel, and zinc are found in wastewater, primarily from dish and laundry detergents, bleach, disinfectants, and degreasers. Alternative products that reduce the health hazards for workers and building occupants, as well as reduce the impacts on wastewater treatment and groundwater, can easily be procured and are effective.

**Green Cleaners.** Environmentally friendly cleaning agents have fewer toxic ingredients and generally do not contain phosphates, chemical compounds that are harsh on wastewater treatment and can cause problems for the bodies of water that receive wastewater effluent. Environmentally friendly floor waxes do not contain zinc and other heavy metals that interfere with the sewage treatment process. Manufacturer claims of environmentally friendliness should be viewed with some skepticism. Environmental leaders and maintenance managers can consider checking the ingredients or verifying the product with a third party such as Green Seal.[36]

Although not usually considered for institutional applications, basic cleaning products such as soap flakes, Fels Naptha soap, baking soda, and Bon Ami are often effective cleaning alternatives that can be used on a large scale.[37] A solution of vinegar and water is an effective window cleaner. Baking soda can substitute for some more abrasive cleansers and is nontoxic. Baking soda can also be used to "sandblast" outside surfaces of buildings to remove graffiti and other stains without damaging the surfaces. Buildings and grounds departments can try these and other environmentally friendly alternatives to gauge their effectiveness. These products can become the standard for routine cleaning, with more toxic and aggressive cleaners resorted to only when the others fail. Pilot-testing products before making a university-wide switch is essential, and some products will require employee training in their use.

Determining a product's toxicity and finding effective alternatives can be time-consuming. Materials Safety Data Sheets (MSDS), available from manufacturers, give basic information about products' toxicity and main ingredients. Although an MSDS does not necessarily list all ingredients or provide information about the environmental effects of the product, it is a good place to start.[38] Cleaning products that do not contain butyl cellusolve or petroleum solvents offer reduced threats to workers and the environment. Other chemicals to avoid are ethylene glycol, formaldehyde, hydrochloric acid, hydrofluoric acid, kerosene, phenol, potassium hydroxide, sodium hydroxide, sodium hypochlorite, sodium metasilicate, and sulfuric acid.[39] Look for products with moderate pH, without carcinogens, and nonreactive products. Products with the minimum amount of fragrance and dye are usually less toxic, and these added substances have little cleaning value. Avoid products that demonstrate aquatic toxicity as indicated by warning labels. Phosphates are common additives to detergents and can cause algae blooms in water bodies that receive wastewater. Ozone-depleting chemicals such as solvents should also be avoided; water-based cleaners are generally more benign.[40]

**Central Mixing.** Many institutional cleaning products come in concentrated form and are diluted for use. Concentrated cleaning products generate less waste than ready-to-use solutions and are often less expensive on a per job basis, but their effectiveness and caustic properties vary according to their dilution rates. The Tufts manager of custodian services described his staff's method of mixing concentrated solutions as "the glug-glug method," whereby the dilute mixture is created by hand and results in random concentrations. Greater reductions in the quantity of cleaners used are achieved when custodial staff measure the cleaning solutions accurately. Central mixing and dispensing units help avoid a great deal of waste and often pay for themselves rapidly. Ithaca College reduced its cleaning costs by 80 percent by switching to a concentrated cleaner that came with built-in measuring spouts.

## Hazardous Waste in Buildings and Grounds

A university's waste stream may contain regulated components or materials that should be considered hazardous. Although many of these are in the university laboratories (see chapter 7), B&G also handles hazardous materials, both regulated and unregulated—for example, lead, used motor oil, batteries, copy machine and laser printer toners, tires, and chemicals.

### Recycle Automotive Wastes

B&G departments must properly handle motor oil and other vehicle maintenance items. This requires constant employee training and new systems to manage these wastes. Used oil can be re-refined, as can gasoline, jet fuel, heating oil, and lubricating oil. If these wastes are disposed of improperly, just a few drops can pollute groundwater or foul wastewater treatment systems. Car and truck batteries should always be recycled. Used tires can be retreaded or sold to a reclamation facility that shreds them for specialty surfaces. The surface of the indoor track at Bates College is made from old tires.

Buildings and grounds departments must collect automotive wastes, segregate each waste, and label each carefully. Mixing wastes can be costly; for example, a pint of brake cleaner can turn 2,500 gallons of used oil from an exempt recyclable waste into a hazardous waste. Wastes should be stored above ground on an impervious surface and tanks checked regularly for corrosion or leakage. The university environment, health, and safety department can assist B&G in securing a responsible waste hauler or recycler. On-site or mobile recycling units can also be used to recycle antifreeze and oil, cutting disposal costs. Spent degreasing solvents can be recycled on- or off-site.[41] Hazardous wastes such as solvents or paints should never be burned on-site without proper facilities, training, oversight, and permits.

**Minimize Release of Asbestos Fibers**

Asbestos was widely used from the early 1900s into the 1970s as insulation, fireproofing material, and acoustical surfacing material. These uses were banned by the EPA in the mid-1970s because of mounting evidence that exposure to airborne asbestos fibers causes lung damage and cancer. It is a common misconception that the presence of asbestos is hazardous to building occupants. In fact, the mere presence of the material is not; rather, exposure to asbestos fibers is hazardous. EPA advises that the best way to reduce asbestos exposure is often to leave it in place because improper removal can create a dangerous level of exposure where none existed previously.[42] Federal regulations, the Asbestos Ban and Phase-down Rule of July 1989, banned the production of most asbestos-containing products (the use of asbestos in brake pads is one exception) in the United States by January 1996 but does not require the removal of these materials where they are currently in place.

Since intact and undisturbed asbestos does not pose a health risk and proper maintenance of the material can prevent deterioration, an asbestos maintenance program, tailor-made for the facility in question, may be proper management strategy. Such a program begins with an inspection by a qualified professional to determine the locations of the asbestos and establish a complete plan. The program may include repair, encapsulation, enclosure, encasement, or minor removal. Personnel who have access to areas with asbestos should be trained on its locations and instructed to take precautions during their work to avoid disturbing the asbestos. Any evidence of damage should be reported at once. Some inspectors recommend routine evaluation of dust to ensure that it does not contain asbestos fibers.

When a building is being renovated or the asbestos material has been damaged beyond repair, removal of the asbestos is required. An asbestos removal professional should undertake the job in accordance with federal regulations that require a "zero visible emissions" standard under Clean Air Act regulations by containing the material so it does not enter the rest of the building and ensuring the health of the abatement workers. State laws governing asbestos removal may be more stringent than federal laws regarding threshold levels of materials and may present differing time frames where the standards apply.[43]

### Monitor and Replace Leaking Underground Storage Tanks

Underground tanks store fuel oil, gasoline, and diesel oil on most colleges and university campuses. Leaks from these tanks pose a severe threat to local groundwater and can be costly to the university, in terms of both lost fuel and potential cleanup and liability costs.

Institutions must notify the appropriate state or local agency of every underground storage tank installed after December 1988, specifying the design of the tank, its leak detection system, and the company that installed it. Tanks must be fiberglass-reinforced plastic and have appropriate anticorrosion protection. The tank must be installed according to the manufacturer's instructions and have a release detection system that may include tank tightness testing, manual gauging, or other monitoring. Older tanks that do not meet new tank standards must be upgraded to have an internal lining and leak protection system including spill and overfill prevention. For a complete list of the requirements see Subtitle I of the Resource Conservation and Recovery Act (RCRA) or contact EPA for relevant publications.[44]

### Follow PCB Regulations

Many pieces of equipment, particularly electric transformers but also old hydraulic equipment and capacitors, contain PCBs, which are toxic to humans and other organisms. The Toxic Substances Control Act (TSCA) regulates the use, manufacture, and disposal of these substances. The

EPA, or state agencies in EPA's stead, may inspect academic institutions to ensure compliance with PCB regulations that require regular reporting. Some states may regulate the disposal of PCBs as a hazardous waste.

PCB-containing transformers, defined as those containing PCB in concentrations greater than 500 parts per million (ppm), may be used for the rest of their useful life, but they will require safety inspections, EPA-approved labeling, and registration with the local fire department. They also must be inspected every three months for leakage. If leaks are found, the leak must be stopped and the transformer contained and inspected daily until the leak is repaired. Spills must be cleaned up within forty-eight hours. Detailed records of transformer servicing and inspection are required; fines of up to $25,000 per day can result for improper record keeping. B&G managers should be aware of PCB-containing equipment on campus and ensure that it is handled properly.

**Use and Dispose of Paints Properly**

Paint, varnish, stain, paint thinner, and paint stripper are hazardous materials and should be handled and disposed of as such. Exposure to heavy metals and solvents are of greatest concern during paint application and may cause dizziness and drowsiness. With prolonged exposure, damage to vital organs can occur, so B&G departments should ensure that waste reduction is the first step in a paint management program.

Waste reduction begins by ordering only what is needed. Painters using spray guns should work only with those with high volume and fully enclosed low-pressure guns. Smaller spray gun paint cups can also reduce waste by limiting the amount of leftover paint and the need for solvents. Spray techniques can reduce paint waste—for example, by setting the correct air pressure and minimizing the overlap with each stroke. Painters who are spraying paint and primer or preparing surfaces by sanding or grinding paint should always wear respirators. Buildings and grounds managers should see that these practices are followed in order to reduce liability, waste, and the use of hazardous solvents on campus.

Selecting less hazardous paints (low-volatile organic compounds) and paint-related products such as latex rather than oil-based paints reduces the hazard on- and off-site. Low-solvent, water-based paints and even more traditional paints, such as milk-based paints, are readily available from major paint manufacturers.[45] Using mechanical stripping methods like sandblasting or bead blasting or nonphenolic strippers can also reduce risks. Keeping paint and thinner cans closed and protected from temperature extremes prevents waste.

Waste oil-based paint, paint thinner, and turpentine must be disposed of as hazardous wastes. Keeping paint and thinner separated makes them easier to recycle. The environmental health and safety department (or other department in charge of waste disposal) can help B&G store paint products appropriately and ensure that wastes are picked up with regular hazardous waste.

## The Clean Air Act

The Clean Air Act of 1990 regulates large air emissions from factories and utility companies. Some university systems may be regulated too, such as emissions from boilers and on-site incinerators, air-conditioners and refrigeration units, and transportation plans (see chapter 6). Of particular concern are air pollutants such as nitrous oxides (NOx) and volatile organic compounds (VOCs), precursors to ground-level ozone. There are different requirements regulating the emission of ozone precursors depending on the institution's location. EPA has designated levels of "nonattainment" for various regions and links compliance requirements to those levels.

Humans, animals, and ecosystems rely on the earth's protective ozone layer to screen out harmful ultraviolet rays. Twenty years ago, chlorofluorocarbons (CFCs) were found to destroy this ozone layer. Recognizing the need to curb the release of compounds that deplete atmospheric ozone, the Clean Air Act (sections 604–606 and 608–609) regulates the phase-out of the production of ozone-depleting chemicals, CFCs, and hydrochlorofluorocarbons (HCFCs), specifically HCFC-22 and HCFC-123, and mandates the reclamation and recycling of these chemicals. Regional EPA offices or state air management offices can assist with the specific regulations. EPA's stratospheric Ozone Protection Hotline (800/296-1996) is also a good source of information and can provide specific listings of ozone-depleting refrigerants and their substitutes.

## Evaluate Alternative Cooling Technologies

CFCs are the most common heat exchange agents in refrigeration and air-conditioning. Because they will be phased out of production, they may now be unavailable to recharge existing systems. The same will be true for some HCFCs in 2010 and 2020. CFC compounds that will be available will have been recycled or even imported illegally. As a result, the cost of recharging units with CFCS will be increasingly expensive. Thus, the best long-term way to reduce the university's dependence on ozone-depleting chemicals and the regulations that come with them is to find alternatives to their use. Traditional vapor compression and absorption technologies can use ammonia, water, or hydrocarbons as refrigerants. Evaporative cooling, zeolite cooling, and thermoelectric cooling are in use or are in development.[46] Engineering students can probably provide B&G managers with valuable research about the alternatives to ozone-depleting compounds on their campus.

Replacing or upgrading the chillers in air-conditioning units that produce cooling for medium and large buildings can eliminate the use of CFCs or dramatically reduce their loss into the atmosphere by decreasing losses from as much as 15 percent of the annual charge to less than 0.5 percent.[47] In addition, new chillers use 25 to 30 percent less electricity than the units they replace at both full and partial load conditions.[48] When chiller replacement is coupled with efforts to install energy-efficient lighting and efficient office equipment, the reduction in waste heat may allow the replacement unit to be smaller and consequently less expensive to purchase and operate than its predecessor. Replacing a chiller can provide an opportunity to use fuels that are cleaner (and often less expensive) than electricity, such as natural gas.

## Monitor Refrigerants and Fix Leaks

The law prohibits individuals from knowingly venting ozone-depleting compounds, used as refrigerants, into the atmosphere while maintaining servicing, repairing, or disposing of air-conditioning or refrigeration equipment. Leaks are the primary cause of refrigerant loss to the environment, and most losses are attributable to improper operation of equipment. Low-loss fittings and valves and high-efficiency purge devices can significantly reduce refrigerant losses during operation. Maintaining accurate operating logs—for example, by using computerized building management systems—can help B&G managers to document refrigerant levels and rapidly detect leaks.[49] If the university owns equipment that has more than fifty pounds of refrigerant, the law requires accurate servicing records documenting the date and type of service, as well as the quantity of refrigerant added. To determine the annual leak rate, the EPA suggests keeping service receipts that contractors must provide, regardless of the size of the operation. If university equipment has a leakage rate over 15 percent of charge capacity, the university is legally responsible for complying with the leak-repair provision that requires repair of substantial leaks within thirty days or the implementation of a retrofit plan within one year.

Broken and outdated air-conditioning and refrigeration equipment must have their refrigerants reclaimed before they can be disposed of or recycled. Universities must use certified technicians with certified equipment for this process. Although small appliances such as household refrigerators and air-conditioners are exempt from this requirement, the green university will ensure that the refrigerant from even these small sources is reclaimed.

In order to avoid releases of ozone-depleting compounds, technicians must evacuate air-conditioning and refrigeration equipment to established vacuum levels. They must also use recovery and recycling equipment that is certified by an EPA-approved equipment testing organization. A university with on-site technicians who service the equipment must see that the technicians observe these no-vent and recycling regulations, ensure that the equipment is certified, and help technicians themselves become certified. If the university contracts for the service of this equipment, it should ensure that the contractors employed are using certified equipment and are certified themselves.

University personnel who maintain university vehicles should also use registered technicians certified under section 609 of the Clean Air Act to maintain vehicle air-conditioning equipment.

**Maintain Boilers in Compliance with the Clean Air Act**

In parts of the country with poorer air quality or in universities with larger heating plants, the Clean Air Act may require special pollution controls or permitting. Regulations are changing, and state regulations may also apply, so B&G or energy managers should check with their regional office of EPA and state environmental protection office for accurate local requirements. Local utility companies can also assist universities in determining if these requirements affect them.

The objective of the Clean Air Act is to lower the quantity of air pollutants from major sources by requiring the uses of the best available control technology (BACT) to reach the lowest achievable emissions rate (LAER). Sources required to use BACT emit more than 50 tons per year of VOCs or NOx or have an output of at least 50 million btu per hour. This probably does not include any but the largest university boilers or generators. The law also regulates emissions of sulfur dioxide, but again these regulations apply only to the largest generators. Before purchasing a new boiler, B&G should check to make sure the boiler's emissions will comply with Clean Air Act regulations.

Universities that operate units under the EPA thresholds should nevertheless maintain their boilers to burn as cleanly as possible. Faculty and student researchers can work with B&G to evaluate the emissions and determine how pollution controls may reduce emissions from boilers, onsite incinerators, cogeneration facilities, and generators.

### Avoid Ozone-Depleting Fire Extinguishers

Halons are chemicals that have been widely used in fire extinguishers since the 1970s. Halons 1211, 2402, and 1301 are the most commonly used in firefighting; halon 1301 is about ten times more potent an ozone destroyer than CFC-11.[50] These halons are regulated as Class One CFCs under Title VI of the Clean Air Act. This means that their production was banned in 1993, so although their use remains widespread, recharging them will become increasingly costly and difficult.

Carbon dioxide, foam, and powder are alternatives to halons for extinguishing agents. Water extinguishers are also alternatives (although their use in university dormitories often invites high-power water fights). Some of these alternate extinguishing agents require some modifications to existing equipment, especially in fixed, rather than portable, systems.

## Pest Management

Universities harbor a range of pests: insects such as moths and cockroaches, rodents such as mice and rats, and larger mammals such as raccoons. These pests are often a nuisance, sometimes a health hazard, and although they may have some advocates, they are generally considered unpleasant and undesirable. Pesticides and rodenticides are chemicals that receive widespread use for removal of pests, but their repeated use can cause them to be ineffective. Further, pesticides are dangerous to the health of building occupants and applicators. The hazards or pesticides are often understated by their MSDS since these reports account for the active ingredient and ignore the inert ingredients, despite the fact that these inert ingredients may not be allowed as active ingredients.[51]

Integrated pest management (IPM) uses information about the life cycles of pests and their habitats to manage pest damage with the least hazard to people, property, and the environment. IPM techniques do not necessarily eliminate the need for chemical pesticides, but they limit their use.

**Ask Students and Staff to Help with Basic Housekeeping**

A successful IPM relies on the cooperation of all members of the university community. When introducing an IPM program, B&G can discuss its objectives and the simple steps that students and staff can take to reduce their personal exposure to pesticides. Because pests seek environments that are conducive to their needs for food, water, and shelter, students have a special role in successful IPM programs by disposing of food properly and keeping rooms free of crumbs, rotting food, and other items that attract pests. Staff, particularly in kitchens, offices, and other areas with food, also have a role in this effort.

**Set Pest Management Objectives and Action Thresholds**

In order to undertake pest management most effectively, B&G needs to work with affected departments to determine the objectives of the pest management plan. This plan should include thresholds, or a bottom line of acceptable action and results. Pest management objectives and thresholds may vary depending on the campus location. For example, small numbers of rodents may be acceptable around central solid waste collection facilities but not in dining halls. Following are examples of pest management objectives:

- Manage pests that may occur on-site to prevent interference with the learning environment.

- Eliminate injury to students and staff (e.g., from rodent bites or disease).
- Preserve the integrity of the buildings and structures (e.g., from ants or termites).
- Provide the safest athletic surfaces possible (e.g., avoid moles or animal holes).
- Protect human health.
- Ensure appropriate sanitation.[52]

**Identify and Monitor Pests and Their Sources**

Routine inspections of facilities to identify pests are essential to successful pest management. Inspections should be detailed and involve looking into cracks (in walls), crevices (computer keyboards), cupboards, and trash areas for the offending pests (roaches, ants, mice, etc.). Staff and student interviews can help determine the types of pests that are prevalent and their habitats. Evidence of pests, such as droppings, chewed wood, or holes in food containers, are also important to record.

**Apply Pesticides Only Where Needed**

Successful IPM programs rely on identifying pests and their habitats and applying basic sanitation and physical controls such as screens and doors, and mechanical controls such as traps. Table 3.3 summarizes the IPM strategies that can be effective for managing common pests found at colleges and universities.

When prevention and physical barriers are ineffective or only partially effective, limited and targeted pesticide use may be warranted. Because of the toxic effects of pesticides, they should be applied only by licensed professionals and for their intended use. B&G should limit the use of sprays and fogs and use direct applications in cracks and crevices instead. When possible, apply pesticides only when buildings are unoccupied and when food is stored securely to prevent any exposure.

On college grounds, keeping grass healthy will reduce the need for pesticides and herbicides. When necessary, consider the use of dry pesticides that are spread on the ground and then watered to reduce the airborne exposure that results from spraying. Insecticide soaps or oils applied at targeted times can interrupt the life of a specific pest. Connecticut College reports having avoided the use of pesticides or herbicides on the college green for five years by planting only native and pest-resistant species of plants.[53]

On some campuses, B&G managers find that they can dramatically reduce the use of pesticides by changing their pesticide application rates to address problems as they arise (these problems should be fewer and further between with the use of IPM) rather than contracting for regularly scheduled spraying several times each year. B&G should rely on regular inspection of the facilities and grounds by B&G staff, rather than by the pesticide applicator to set a pesticide application schedule.

MSDS for all pesticides used on campus by university staff or outside contractors must be on file and available to the university community. Laws that govern the access to information about chemicals in the workplace, often called right-to-know laws, also apply to university staff who use or are in contact with pesticides.

## Grounds

The grounds department is responsible for the maintenance and improvement of university grounds, including athletic fields. As in other departments, some grounds work is carried out by university staff, and other work, such as tree work, pesticide applications, and fertilizer applications, are often contracted to outside companies. In some climates, the grounds crews are responsible for snow removal and leaf raking and removal. When the university holds special events, the grounds department is often pressed into overtime service to spruce up the campus. These varied responsibilities offer a range of environmental stewardship opportunities in the maintenance of university grounds.

### Evaluate Standards for Lawns and Athletic Fields

The lawn care industry has defined the ideal lawn as composed of grass species only and free of weeds and pests. It is continuously green and mowed low and evenly.[54] This lawn is the same in most climates, even the most arid areas. The green university, however, does not necessarily look like a country club, with rolling green lawns and fields, perfectly manicured and watered. Instead the environmentally sensitive university grounds manager will be willing to let some spaces fill with indigenous

**Table 3.3**
Integrated pest management strategies

| Typical pests | Locations | Strategies |
|---|---|---|
| Indoor | | |
| Mice, rats, cockroaches, ants, flies, wasps, hornets, yellow jackets, spiders, termites, carpenter ants | Entryways: doorways, windows, holes in exterior walls, openings around pipes, ducts | Keep doors shut<br>Install weather stripping<br>Caulk cracks<br>Repair screens<br>Keep vegetation 1 ft. from structures |
| | Classrooms and offices, including laboratories, gymnasiums, and hallways | Allow food only in designated areas<br>Keep plants healthy<br>Keep area dry<br>Remove standing water<br>Store food in sealed containers<br>Clean animal cages regularly<br>Clean lockers routinely<br>Vacuum carpets frequently |
| | Food preparation and serving areas | Store food in inaccessible containers<br>Store waste in secure containers<br>Remove waste at end of day<br>Place screens on vents, windows, and floor drains<br>Remove all food sources (crumbs, water leaks, etc.)<br>Avoid grease buildup<br>Use traps or glue traps |
| | Rooms with plumbing | Repair leaks promptly<br>Clean floor drains |

| | | |
|---|---|---|
| | | Seal pipe chases<br>Keep areas dry<br>Store paper products in dry areas off the floor |
| | Maintenance areas | Promptly clean mops and mob buckets<br>Dry mops by hanging<br>Allow eating in designated areas only<br>Clean trash cans regularly |
| Outdoor | | |
| Mice and rats | Parking lots, loading docks, and dumpsters | Clean trash containers regularly<br>Secure lids on trash containers |
| Beetle grubs, sod web-worms, moles, diseases | Turf such as lawns and athletic fields | Maintain healthy turf<br>Raise mowing height to enhance competition with weeds<br>Vary mowing pattern to reduce compaction<br>Let soil dry between waterings<br>Provide good drainage<br>Leave clippings on turf<br>Dethatch the turf<br>Apply fertilizer only as needed and according to soil tests |
| Aphids, Japanese beetles | Ornamental shrubs and trees | Apply fertilizer and nutrients during active growing season<br>Prune branches to improve plants and prevent access to structures<br>Use pest-resistant varieties (check with local extension service)<br>Remove diseased plants |

Source: Environmental Protection Agency, *Pest Control in the School Environment: Adopting Integrated Pest Management* (Washington, DC: U.S. Government Printing Office, 1993).

wildflowers and allow broadleaf (often considered to be weeds in the "perfect" lawn) plants on athletic fields.

University lawns and fields receive heavy and demanding use. If it is possible to "rest" a field by limiting its use for a time, the turf will remain healthier, but limiting the use of fields is difficult on most campuses. Increasing the diversity of plants to include clovers, naturally occurring weed plants, and other broadleaf plants can reduce watering needs and increase the resilience of the turf. In addition, turf made up of diverse plants will have a greater resistance to pests and fungus, thereby reducing the need for pesticides and herbicides. This diversification of plants is a natural evolution in existing lawns or fields. Planting diverse species in new lawns will complement this natural process and reduce or eliminate the need to use herbicides and other chemicals.

Changing the standards for lawns and athletic fields on many campuses must be a conscious decision. On some campuses, this decision can be made by the grounds manager; on other campuses, this decision making will need to be done at a higher level. Changes to the commonly accepted standards for a manicured campus need to be accompanied by education of the public and the university administration. For example, when the landscape at the outskirts of a campus is allowed to return to wildflowers, the alumni and trustees may protest that the campus looks shabby unless signs or other material explain the change.

Student or faculty environmental advocates should realize that the performance of grounds managers is customarily evaluated by administrators, alumni, and their peers according to conventional standards. For example, when the director of B&G at a university that considers itself an environmental leader received a letter from an alumnus criticizing him and his staff for maintaining the grounds as pristine as a golf course, he considered the criticism a wonderful compliment! Green universities must reward grounds managers for comprehensive rather than superficial grounds maintenance by supporting innovation and providing education and support for a more natural campus look.

Universities should also strive to avoid the use of sod. Sod farming is an energy-, soil-, and water-intensive industry. Instead, regular maintenance and the selection of appropriate plant species for the climate and region ensure a more environmentally friendly and cost-effective solution to lawn management.

**Reduce the Use of Fertilizers**

Lawn and field maintenance techniques that remove cut grass from lawns eliminate a valuable resource; grass clippings can return organic matter and nutrients to the soil and reduce the need for fertilizer. In universities with large expanses of green grass, leaving mulch in place also saves labor. Mulching lawnmowers are effective at returning these resources to lawns and fields.

Leaf waste, garden clippings, and grass clippings, if they are not returned to the lawn, can be composted to preserve the material as a resource that can be used to add nutrients and water-retaining organic matter to gardens and lawns. Many universities, such as Dartmouth College and Bennington College, compost the college's organic matter on-site; others collect organic material for composting at an off-site location. Some universities combine the leaves with animal waste and bedding and/or food scraps. Composting requires space, labor, and machinery to manage the piles of material. But the resulting material is like "black gold," and returning it to the university's fields and gardens will improve soil quality.

**Reduce Water Use Through Plant Selection and Watering Techniques**

At universities, watering lawns, gardens, and fields is often a major water use. Selecting plants, grasses, and surfaces that are appropriate for the geographical area can vastly decrease the amount of water needed to sustain the plantings. Transferring open grass areas into ground covers, native plants, and shrubs that require little water can reduce irrigation needs and provide a greater amount of habitat for wildlife.

Watering techniques can also affect the amount of water used. For example, watering during the middle of the day increases the amount of water lost to evaporation, both during the application of water and as it is absorbed into the soil. Watering in the early morning can decrease this water loss. Evening watering can also decrease water loss but may leave grasses and plants damp and prone to disease. In many universities, night and early morning watering is problematic because grounds crews are generally off duty during the evening hours, but water timers and other devices can solve this problem. Watering equipment such as sprinklers and water cannons also contribute to the water lost by evaporation. In many places irrigation systems and soak hoses are cost-effective alternatives. Water should be applied deeply and less frequently to encourage deep root growth; however, only one inch of water is needed to wet the soil to a depth of four to six inches. Water reaching the grass or gardens can be measured with a rain gauge or a shallow can.

### Prevent Erosion

Wind, flowing water, and human activity cause soil loss and erosion. On a global scale, soil loss is a major problem, with about 7 percent of the world's topsoil being lost from potential cropland each decade.[55] Soil erosion causes the loss of nutrient-rich topsoil and pollutes rivers and other waterways with pesticides, fertilizers, and sediments; soil in municipal sewer systems can complicate the treatment of wastewater. On university campuses and university-owned lands, appropriate plantings of grasses, shrubs, and other vegetation can help prevent the loss of soil, especially on hillsides and steep slopes. During campus construction projects in particular, care must be taken to prevent the runoff of soil into roadways, culverts, and watercourses from excavation sites and newly installed topsoil. Hay bales and siltation fencing are two effective techniques. University grounds or construction managers must enforce this valve with outside contractors who may not share the university's concerns. When runoff from construction sites is severe, the university may be in violation of local ordinances regulating erosion and stormwater discharges.

### Evaluate Snow and Ice Management Techniques

Snow and ice create hazardous driving and walking conditions. In these cases, grounds managers must balance the need to provide safe and secure walkways, parking lots, and roadways with the damage to plants, gardens, and trees caused by salt. In turn, the damage from salt is balanced against the wear and tear on carpets and floors caused by sand. Salt, moreover, can increase the sodium content of water supplies and damage waterbodies, and the humans and wildlife that use them. Runoff from stockpiles of salt and piles of salt-laden snow can intensify the problem.

Reducing salt use, finding alternatives to salt, and careful application of salt can all help to reduce the damage to plants and waterways. Grounds managers might maintain separate stores and trucks for sand and salt application to limit use. Sand can often be substituted for salt on flat roads and the parking lots. Salt is ineffective at very cold temperatures, so it should not be used in those cases. Alternatives to salt are very expensive, but future research will make these nonpolluting alternatives more attractive. Around sensitive gardens and trees, these alternative chemicals may be cost-effective in the long run since they reduce plant loss and damage.

**Involve Students and Staff as Stewards of the Landscape**

Students and staff are ready resources to serve as stewards of a garden, field, tree, or shrub. In the same way that Adopt a Highway programs have been successful, college community stewardship programs can help grounds crews pick up litter, care for plants, and otherwise beautify the campus. Students can team with B&G crews to serve as stewards for a corner of the university grounds. Often these stewardship efforts are undertaken by individuals on their own initiative. At Tufts, a professor transplanted flowering shrubs that were about to be destroyed by a construction project, to a safe location, and a dining worker watered the flowers daily while the kitchen he ordinarily worked in was vacant during renovation.

**Substitute Nontoxic Products for Toxic Products Where Possible**

Steps should be taken to reduce the toxicity of products whenever possible. The grounds department can make choices about pesticides, fertilizers, and herbicides. As with cleaning supplies, identifying these alternatives can be time-consuming, but often the alternatives result in a lower overall cost. Many schools have successful switched to water-based paint for lining the boundaries of the athletic fields in order to reduce the use of hazardous substances. Organic alternatives to herbicides, pesticides, and insecticides can be found by contacting a local chapter of the Audubon Society or the local extension service. And students can be a ready resource for grounds managers who want to investigate and test alternative products.

## Building Construction and Renovation

Constructing new buildings and renovating existing ones provide the university with significant opportunities to incorporate energy efficiency, recycling, healthy indoor air, water conservation, sustainable products (e.g., wood products produced from wood that is grown sustainably), and new fuels and technologies into building design. Renovations using state-of-the-art design methods and construction materials are sometimes more costly but save in operating costs throughout a building's life. In fact, the Natural Resources Defense Council (NRDC) building in New York City saved $2.30 per square foot in the construction over typical construction costs while incorporating energy-efficient technologies and nontoxic building materials into their project. Its energy improvements alone had an estimated payback of 5.5 years.[56]

The planning and design of Oberlin College's planned Environmental Studies Center goes several steps beyond efficiency and the use of low-impact materials. Design criteria for this innovative building reflect an understanding that the physical environment of buildings and landscapes offers a multitude of learning opportunities. There are important design criteria for this facility that include:

- The project will incorporate the building and landscape design together.
- Building and landscapes systems will work with natural energy systems.
- Materials will come from local sources to the highest degree possible.
- The building will be a net energy exporter.
- The landscape will teach and communicate about the site and region.
- The maintenance of the landscape will be participatory.
- The landscape design will be a seed for the future development of the campus and region.
- Living systems will be used for treating wastes.
- The project will be a powerful, memorable statement of the art and ecology of design.[57]

Clearly this project takes a forward-thinking approach to new design and construction that can be a model for others.

The new construction and renovation processes encompass these areas:

- Assessing the costs
- Determining the program needs for the department that will use the building
- Identifying the site location
- Selecting an architect and building design
- The building process itself

The process typically involves administrators and building occupants plus the university's construction department, usually a division of B&G, responsible for translating the needs for a new building into an actual building. Once the initial needs are determined, the construction department oversees the design, fees, capital expenditures, survey team, and architectural firm. It also works with the administration to present the plan to the trustees for funding (in the case of large projects).

Construction departments serve as the planners and coordinators of this process. Their mission is to satisfy the end users, usually an academic or administrative department, without spending too much money. They must also ensure that a new or renovated building fits with the other campus buildings architecturally and operationally and conduct the project in compliance with the many regulations that affect building construction, utilities, and the environment. A construction department will also be concerned that newly constructed spaces or buildings use systems that the university is equipped to operate and maintain.

The green university will design its buildings so they meet programmatic needs—classroom, library, study space, or living space—and operational and environmental needs. Today there is ample evidence that both can be achieved with success. Often the biggest barrier to implementing new and efficient technologies is that these technologies are unfamiliar to maintenance staff and university trades people (e.g., plumbers and electricians).

There are many design concepts and methods to create climate-responsive and energy-efficient architecture that can be used in new construction or renovation. These techniques can result in energy costs that are 50 to 90 percent lower than those of conventional buildings.[58] In addition, these green buildings can improve working conditions and productivity.[59] For example, the University of Chicago's Graduate School of Business designed and constructed a new building that reduced the annual energy cost from $698,000 using traditional technologies to $300,000 using new technology.[60]

**Select an Architect Familiar with New Technologies**

To design the most efficient and healthy buildings, a university must work with architects who are advocates for new and efficient technologies. Construction managers should look for an architectural team that understands how building components work together and complement each other and has designed buildings that use new technologies.[61] As the construction team describes its vision for the new building or renovated space, it might consider how the building will operate and include the energy and health aspects as the building is being sited and planned. Dartmouth College's Energy Council's New Construction Energy Standards advise, "Architects, engineers and other professional consultants should be evaluated in light of their successful experience in energy conservation in addition to other selection criteria. Consultants should be impressed strongly with Dartmouth's concern . . . with a project's total life cycle cost in addition to the first cost of construction."

Architects are often paid a percentage of the building's construction cost—an incentive to install larger systems, such as HVAC systems, than may be needed and to overbuild many other aspects of a project. By the same token, architectural firms are often held accountable if building systems are improperly sized or fail to work as promised—another incentive to overbuild or to stick with old, reliable technology. Construction managers should try to compensate architects in ways that provide the right incentives for efficient and correctly sized technologies.

University buildings that are models for green design and efficiency, such as Oberlin's Environmental Studies Center, involve a design team that is familiar with and committed to the use of methods and technologies that are not necessarily routine. Sometimes university environmental leaders will need to push hard to ensure that the most familiar technologies and design are not plugged in. Architects may be fearful that these technologies are more costly, but this need not be the case. The University of Northern Iowa built its Center for Energy and Environmental Education for $104 per square foot, a reasonable and market-competitive rate. This building operates with annual energy savings of about $9,000.[62]

**Select an Appropriate Site and Design**

The placement of a building can influence how it uses sunlight or is shielded from harsh winds. Roof overhangs, trees, hillsides, and other buildings provide shade, erosion control, and protection from the wind. Topography and soil conditions can affect temperature in a building by channeling winds and moisture. Site conditions can have an impact on the views inside the building, as well as on the noise from wind or traffic. Proper siting of a building, combined with insulating materials, can reduce the operating costs and the construction costs for heating and cooling equipment. The NRDC's new building's cooling load (the energy needed to cool the building) was about 300 kBtu per hour as compared with 609 kBtu per hour for a standard building, thereby reducing peak cooling load (the highest load, usually midday, during the time when a utility's demand for power is highest and thus most costly) by 50 percent.[63] Depending on the region of the country, buildings should be sited to maximize winter and minimize summer solar heat gain. This is accomplished on a site-by-site basis, so construction managers should discuss siting with architects early in the planning process. Building siting can also maximize the use of natural light in order to reduce the need for artificial lighting and corresponding electricity demand.

**Incorporate Aggressive Environmental Criteria into Construction Standards**

Most institutions have construction standards that dictate the needs and expectations of the university to contractors, architects, and other building professionals who are undertaking new construction or major renovations. These standards specify the minimum requirements for university buildings and are designed to ensure uniformity of equipment, techniques, and quality.

Many colleges and universities specify insulation, lighting, and HVAC criteria for new projects, but few have realized the opportunity to make all new and renovated university buildings highly efficient and healthy spaces. To do this, universities must go far beyond complying with building codes. Sometimes they will need to trade a lower initial cost for the lowest combined installation and operating costs. Construction standards are an effective place to incorporate environmental criteria such as efficiency, attention to recycling, prohibitions on the use of rare wood products, the use of recycled building products, or the use of nontoxic paints.

When constructing a new building there are sometimes conflicts between maximizing efficiency and how the building will be used. For example doors propped open can disturb temperature balances or curriculum may itself be energy intensive. Energy-conscious construction managers find a conflict between the laboratory-intensive educational program and the associated fume hoods that vent large quantities of heated or cooled air, and the desire to have an energy-efficient science building that reduces air losses.

**Design an Efficient Building Envelope**

The building's envelope or shell—its walls, ceilings, roofs, doors, and windows—determines how internal and external forces affect the comfort and efficiency of the building. The efficiency of the envelope is affected by its design and construction, as well as the materials used as insulation, glass, seals, and air locks.

Building envelope considerations are often sacrificed for design considerations, such as using glass that is standard double pane rather than windows with greater insulating properties, such as triple-pane windows or windows with reflective coatings. The building envelope design may incorporate a buffer space from solar gains or thermal losses to critical working or living spaces. The Campus Information Center at Rensselaer Polytechnic Institute uses a buffered sun space that heats to above the required temperature. Heat is pumped from this space when needed or is vented to cool the space.[64]

**Design for Recycling and Waste Reduction**

The lack of safe, clean, and dedicated space for the collection and storage of recyclables may be one of the biggest barriers to effective recycling programs. Likewise, poorly laid out spaces or distant storage facilities may thwart waste reduction. The design of a new building can incorporate these considerations from the start. Including the recycling manager or an environmental spokesperson in the building design process can help ensure that these issues are not put aside. At the University of Colorado, for example, construction projects must get recycling office approval.[65] Provisions for source separation of construction and demolition material for recycling and reuse should also be included in any construction project. The National Audubon Society's building incorporated chutes to a central storage area that allow each floor to dispose of paper, aluminum, plastics, and food wastes. Other materials are collected on each floor, and each desk has a three-slot waste bin. Wastes are processed and marketed from a central recycling center and biodegradable material is composted on-site. Audubon's goal is to recycle 80 percent of the waste material that enters the organization.

Careful planning of supplies and an aggressive program to capture scrap for reuse by university departments such as drama and art can reduce waste from renovations and new construction. Recycled products can be incorporated into paving, asphalt, site furnishings, pipes, and lumber (see chapter 4).[66]

**Use Efficient Lighting**

Lights are often integral to the architectural design, and construction managers can ensure that they are highly efficient at the design stages. Architects should be expected to minimize decorative lighting that highlights features but provides little actual light. Too often institutions that are rapidly updating their lighting in existing buildings allow architects to specify the same less-efficient fixtures for new spaces that are being removed in other locations. Construction managers should prohibit the use of incandescent lights. Spaces should be designed so that task lights are used rarely or used only in areas where the surrounding ambient light levels are low. Dark colors for walls and ceilings should be avoided since darker colors absorb light, thereby requiring additional lighting to reach comfortable light levels. Whenever possible, natural light should be used and daylight sensors installed to shut off lights automatically when daylight levels are adequate. More light is not necessarily better light. Construction standards can specify the light levels that are needed and require that the design meet and maintain, but not exceed, those levels. Standards such as those provided by ASHRAE can provide updated guidance for construction projects.

### Consider Environmentally Friendly Technologies

Given that universities are home to researchers and scientists working to document environmental problems such as climate change, the greenhouse effect, ozone depletion, and air pollution, it is amazing that so much of this research takes place in conventional, and often inefficient, facilities. New energy-efficient building materials, insulation, windows, air-handling systems, and HVAC systems can make building and operating buildings more cost-effective. Solar collection and storage are important aspects of building design. Direct cooling from natural ventilation, pressure differentials, and indirect cooling where air is pushed or pulled through the building can dramatically reduce or even eliminate air-conditioning needs. Window selection, window films, shades, and draperies can affect building comfort and efficiency. The Comstock Building in Pittsburgh was built in 1983 using energy-efficient lights, HVAC, and windows in 175,000 square feet of space designed to maximize heat and cooling retention. The savings from the construction exceeded $500,000, and energy consumption and operating costs are about half that of other large office buildings in the area.[67] In NRDC's building, double-paned, low-emissivity windows provide nearly the same insulating value as conventional walls and about five times the insulation of ordinary windows. "Smart windows" are now available that automatically darken or lighten to allow sunlight or reflect heat.

New buildings are perfect opportunities to use new technologies that eliminate the use of CFCs and HCFC, thereby reducing the headaches of monitoring and reporting them. This goal can be accomplished by selecting an alternative cooling technology such as absorption, solar-powered absorption, evaporative, and thermoelectric cooling.[68]

Many other mechanical systems (such as heat exchangers and HVAC systems) are available to make a university's buildings green.[69] Some of these technologies are simple, such as increased solar gain, shading, and natural ventilation. Colleges and universities can look to demonstrate and take advantage of these opportunities. One of the biggest barriers to implementing new technologies is often the reliance on familiar technologies and rule-of-thumb estimations of their effectiveness.

Recycled lumber, recycled paint, recycled carpet, and low-emissive furniture and carpet can also help to make a new or renovated space have less of an environmental impact, reduce the emissions to the indoor air, use fewer resources, and support recycling. Other important products indude formaldehyde-free wood products, sustainably harvested wood, and nontoxic adhesives. (See chapter 4 for more information.) *Environmental Building News* can provide an annotated bibliography of sources that will help construction managers to identify and evaluate product sources.[70]

**Use Alternative Fuels**

Most university systems use electricity, natural gas, and oil to heat and cool buildings, power lights, and heat water, but there are other fuels—among them, cogeneration, solar, and wind—that are more efficient, cleaner, and sometimes cheaper to operate than conventional fuels. These fuels and the systems that run off them are particularly well suited to universities since these institutions usually intend to exist for many years to come and need not yield to pressure for immediate returns on investment.

While not really an alternative fuel, cogeneration produces high-temperature heat and generates electricity. The process maximizes the efficiency of boiler systems and decreases off-site emissions generated by electric power plants. Universities with large central boilers in colder climates can make good use of cogeneration systems.

Solar energy is particularly appropriate for use as a water-heating fuel in applications where large volumes of water are used, such as dormitories, athletic facilities, and kitchens. Even in cold regions, solar hot water heaters and preheaters can effectively raise 40-degree water to a preheated temperature of 80 or 90 degrees on a winter day, as L. L. Bean in Freeport, Maine, demonstrates in one of its office facilities. Solar energy can be cost-effective in these applications. Despite significant initial cost, the savings accrue for years with relatively little maintenance. Heating or preheating water with solar energy is a clean process that does not result in air pollution.

Passive solar heating is usually accomplished by designing the building to capture fully and benefit from the heat of the sun. Passive solar heating should be designed into all new buildings, with attention paid to passive cooling from trees, fans, reflective window films and shades, and new smart window technologies. Passive solar systems are cost-effective and virtually maintenance free (unless manual shades are required). Although passive heating and cooling may not be able to fully heat or cool a building, prudent design can dramatically reduce the demand and thus the size of building heating and cooling systems. The University of Northern Iowa's Center for Energy and Environmental Education uses solar gain from south-facing windows to meet a significant portion of the building's heating load. These same windows are shaded from summer sun by large (2.5 foot) overhangs, thereby protecting the building from overheating.[71]

Photovoltaic systems, which generate electricity from solar power, may not be able to supply the sole source of power for an office building, but they can be effective in lessening the amount of electricity the university must buy, particularly during peak (most expensive) times of the day (usually around noon when the sun is most powerful).

Wind is most commonly used to generate electricity in very small remote applications or in large wind farms. California uses wind to generate enough electricity to meet 16 percent of San Francisco's electricity demand.[72] Windmills on university campuses need sufficient regular winds in excess of 15 mph to operate effectively.

B&G departments are responsible for providing reliable heat, hot water, and other building services and are often reluctant to try unfamiliar technologies. It is important to include maintenance and employee training in budgets for new systems that use alternative technologies, since repair and maintenance may require extra training and equipment.

Although many of these technologies have been around for decades, some still require a great deal of research to improve and refine their operation. The truly green university will find ways to develop and encourage working partnerships between faculty, facilities managers, and sources of research funds. In this way, faculty members with expertise in these areas may be able to pilot-test new technologies or use existing sites for research, while taking advantage of the hands-on expertise of facility managers.

### Consider the Full Cost of Building Operation and Maintenance

Construction managers are often under immense pressure to complete projects within the budget. Consequently, they may sometimes choose to reduce the quality of the initial materials. Despite the facts that maintenance and long-term durability should be major considerations in the construction or renovation of any university space, these costs are often overlooked in construction and renovation. Buildings and grounds departments must coordinate their construction and maintenance managers and budgeting processes in order to realize the savings in operations that new technologies can create. The benefit to the environment will be great as well.

## Special Consideration for Buildings and Grounds Unions

In some institutions, some B&G staff belong to unions. Union contracts may specify who is allowed to perform certain jobs and how these jobs will be done. It is important that environmental initiatives of faculty and students, even if these efforts are limited to data gathering, recognize and follow the rules of the union agreements. For example, a union contract may specify the lighting contractors that the university can select or the extent to which student or faculty can move equipment or perform tasks such as opening lighting fixtures and mechanical rooms in conducting an audit of building systems. The managers of B&G can help in determining if these or other restrictions apply on a campus.

Unions can also motivate environmental actions, particularly as they relate to worker health and safety. For example, a union's concern about the effects of pesticides on university employees who use them may motivate the university's transition to healthier methods of pest management. In contrast, environmental actions can be threatening to unions that may see waste reduction, such as solid waste reduction or the installation of more efficient fluorescent lamps that require less frequent replacement, as actions that threaten jobs.

## Conclusions

The university's buildings and ground department can make significant positive changes in the environmental footprint of the university. Although many of these actions require improved operations and the installation of new and efficient technology, the department's efforts should complement and be coordinated with the efforts of individuals and other departments throughout the campus. Nonetheless, the potential results of B&G's efforts and their operational decisions affect nearly all aspects of environmental stewardship.

The number of action steps outlined in this chapter illustrates the extent and complexity of environmental action that can be taken in the process of building, maintaining, and renovating university buildings and their surrounding grounds. Whether B&G managers and environmental leaders choose a slow, careful, study-it-first approach or a more energetic "just do it" approach, most change will require a strong commitment. This may require new training for employees or new techniques and equipment. Each action needs support, but each is possible and worthwhile.

# 4

# Purchasing

At universities, the purchasing of products and services offers many opportunities to reduce or eliminate adverse environmental consequences. Because these institutions buy large quantities of many items, they can reduce impacts on campus by their purchasing choices, and they can act as advocates for changing markets by demanding environmentally friendly products. Furthermore, some university-wide purchasing decisions that benefit environmental protection can be effective throughout the entire university.

## The Process of Purchasing Products and Services

Although most universities have a central purchasing department, decisions about the purchase of products and services usually occur in three ways. First, *end users*—the faculty, staff, and students who use products or services—specify the products or services they need. In this case, the university purchasing department locates and orders the product and processes the necessary paperwork to ensure that vendors are paid and the appropriate internal university accounts are billed. Here the purchasing department has little influence over the selection of the products or service.

*Central purchasing* decisions, the second type, are made by the purchasing departments or the departments that they run (such as printing services in many universities) about vendors, products, and services. In most universities, central decisions are about products that are used in large quantities or by many members of the university community, such as office supplies and copy paper.

The third type of purchasing decision is the *negotiation of contracts* for university services such as trash removal, painting, and equipment leases for cars, copy machines, and so forth.

Some universities process all orders through a central purchasing department; others have purchasing buyers in each school or department. Still others are entirely decentralized, allowing individual departments to purchase goods and services with little oversight. Regardless of the system, the purchasing department staff and those making purchasing decisions in individual departments are always interested in the cost of products, the product's performance, and the dependability of the supply. University purchasing department staff will enforce university policies, such as those requiring bids from three vendors, when the purchase price of the product exceeds a predetermined dollar value. Overseeing the implementation of appropriate environmental policies can also be the domain of a central purchasing department's staff. Purchasing staff or end users can be influential as well by requesting environmentally friendly products when the specifications for these products are determined, or ensuring that services provided to the university are done with sensitivity to the environment. Sometimes the purchasing department can make major changes in the way business is carried out; in other cases, they can specify commonsense or legal requirements in the hope that contractors or suppliers will be more likely to abide by them. For example, purchasing departments can specify the recycled content of all paper bought by the university or require that painting contractors use latex paints and dispose of waste paint thinner and waste paint properly.

Certainly all environmental problems cannot be solved by the purchasing department alone, but purchasing decisions can make significant progress on reducing solid waste, bolstering markets for recycled materials, reducing energy use, and improving the hazardous materials use practices. To maximize the potential of the purchasing department's environmental action, the department managers and personnel must consider themselves resources, advocates, and sometimes gatekeepers for the rest of the institution. The university can both purchase environmentally sensitive products and promote their use, or even mandate it, within the university.

The Tufts CLEAN! staff found the purchasing department to be open to suggestions but sometimes hesitant or even reluctant to act on the information. There may have been several reasons for this. First, purchasing is generally a pass-through function that processes requests from departments without prejudice or comment. Second, purchasing managers often receive complaints when they undertake seemingly inconsequential changes. The Tufts purchasing department has been effective at reducing waste in central decisions such as copier contracts (more duplexing machines) and in promoting paperless ordering systems. They have hired a student intern to explore the use of recycled products and promote the purchase and use of recycled paper.

Dedicating staff (either formally or informally) to examine opportunities for environmentally friendly purchasing can help make more dramatic progress. For example, Kevin Lyons, the senior buyer at Rutgers University, took it upon himself to change university purchasing contracts to specify a range of environmental objectives. For example, he has included increasing the use of recycled products, changing to products and packaging that are recyclable within university systems, and specifying energy-efficient technologies in university contracts and has saved the university money in the process. He has found creative ways to develop solutions to specific problems by relying on vendors to provide him with information. In contrast, the state of Massachusetts hired a dedicated environmental purchasing coordinator to serve as an advocate and resource for buying recycled and low-impact products through the state office of procurement.[1]

## Selecting Environmentally Friendly Products

The process of selecting environmentally friendly products is not easy or straightforward. Product claims of recyclability, biodegradability, recycled content, or organic status are not regulated in most states and cannot always be trusted. Furthermore, the toxic nature of a product is rarely advertised, and some products that are benign when used have significant environmental, health, or safety consequences in their manufacturing process or off-site disposal. Many alternative products may be better in one area but create other unwanted problems, making the selection of products far from clear-cut or easy. For example, many university food service providers have switched from paper plates and cups to recyclable polystyrene in order to reduce solid waste; however, the manufacture of polystyrene uses styrene, a known carcinogen. Purchasing decisions for environmental stewardship should also be based on campus and regional needs. Using the concepts of waste reduction, energy efficiency, and life-cycle costs presented in chapter 2 provides a general framework, but information about different products is difficult to come by and often incomplete.

Buying only what is needed is the first guideline of environmental stewardship. When information is unavailable, the concept of "less is better" should rule when selecting and using any product. In institutions where the purchasing department largely serves to process orders, policies to curtail excess buying will be difficult. Nonetheless, a purchasing department's environmental commitment can begin with a decision to reduce the volume and weight of material bought by the university.

In order to gain as much information as possible about products, their environmental consequences, and their alternatives, purchasing managers can work directly with interested students to research many products. Research should include the on- and off-site impacts of a product's use, information on its manufacture, waste considerations, and recyclability. Chemistry students can test products, and American studies students can research a manufacturer's relationship with its community. Third-party product watchdogs such as Green Seal can be helpful in identifying products that are harmful or helpful.[2] Articles in environmental magazines and literature may provide product reviews, but one cannot rely on the advertisements in these publications for accurate, unbiased information. Purchasing departments may establish specific goals of waste reduction and efficiency either by declaring a goal or by a participatory process of goal setting. Purchasing department managers can then ask distributors to help make progress toward these goals. Anyone buying any product should feel free to ask for backup information from manufacturers.

In the end, product choices are difficult. Because information is incomplete, customers should reserve the right to change decisions, products, or product criteria as new information is available or as their departments become more knowledgeable.

**Conduct Pilot Tests**

Making changes to the products regularly purchased by the university has some risks, and reliability and familiarity lead people to favor existing products. Purchasing departments can promote more environmentally friendly products and services by pilot-testing them in single departments or buildings to see how they perform. Pilot tests allow the purchasing department to work with product vendors to ensure that delivery, product performance, and recycling or packaging reuse run smoothly. In this way, failed projects can be put right so they do not discourage other campus-wide initiatives.

**Mandate the Use of Environmentally Friendly Products Where Possible**

As appropriate high-performance environmentally friendly products are identified, the purchasing department, a university administrator, an environmental stewardship committee, or other advocates can work to mandate their use. Although the culture at universities usually discourages the use of mandates regarding product purchasing, there are countless universities that require the use of certain products or products from designated companies. For example, most universities have standardized the brand and type of copy paper used throughout the university or may require that departments purchase office supplies from a designated company. This same policy can be applied to the purchase of recycled products or other products with environmental benefit (or less environmental harm).

**Use Budgeting Processes to Reduce Waste**

Individual departments are best able to make choices that reduce their purchases. Budgeting policies that require that departments spend their budgets before the end of the fiscal year or lose the unspent money are necessary accounting procedures, but they encourage consumption. At the very least, administrators should not penalize departments that do not spend their full budgets (e.g., by rampant buying at year end) by cutting their budgets for the following year. Purchasing department programs for reuse and for sharing of equipment of supplies can also reduce consumption.

## Relying on Vendors for Action

To purchase environmentally friendly products and services, there must be a supplier that offers the product. Often, however, a university may wish to maintain a relationship with suppliers that have served the university well over time or have unique products. At other times, the product purchased is not wasteful itself but requires a change in packaging or transportation to reduce its environmental impact. Product vendors, suppliers, and distributors can be helpful in solving these problems for the university. By relying on outside vendors for action, asking the right questions, and including waste reduction, recycling, and efficiency in the standard purchasing criteria, a purchasing department can accomplish significant environmental change without tremendous effort.

In the age of customer service, vendors should be willing to comply with the university's environmental requests. We have all seen how consumer pressure can have a role in forcing environmental action. Well-known examples of the power of consumers include the pressure for "dolphin-safe" tuna and the switch at McDonald's from polystyrene foam to paper-based packaging. Because institutions often buy items in large quantities or many items from a single supplier, their buying power can have great influence on a supplier's willingness to change. Relying on vendors to change their products, product packaging, or environmental service may also help to reduce the environmental impacts at other institutions as well, since vendors will pass these benefits along to other customers.

Box 4.1 shows some of the actions that vendors can undertake on behalf of their university customers. Some vendors will be willing to use their environmental efforts to differentiate themselves from their competition, thereby making themselves more attractive to the university. Waste reduction and recyclability and reduced hazards are the most common vendor efforts, but some vendors may be willing to train university staff about waste reduction or the proper use of their product. Distributors that represent many products can often find products with fewer impacts; for example, a product that is manufactured locally might be substituted for one from across the country to reduce transportation effects. By requesting these products, university purchasers can make changes with

**Box 4.1**
Actions that may be taken by vendors

Waste reduction and recycling
- Ship in bulk packaging.
- Ship in reduced packaging.
- Take back packaging for reuse or recycling.
- Take back used product.
- Offer packaging or product recyclables on campus.

Recycled products
- Paper products with high postconsumer content
- Plastic products, including trash bags, packaging, and lumber
- Building materials
- Reclaimed oil and auto products

Local products
- Locally made or grown
- Locally distributed or packaged

very little in-house effort and also influence other customers who buy from the same vendors and distributors.

Suppliers may act as sources of product information (on the recycled content of a product and its packaging or the toxicity of a product) and technical information such as the best ways to dispose of a resulting waste product. Universities can expect their suppliers to provide information to purchasing managers or students who are researching product choices for environmental impacts or to defend their products to purchasing departments, academic departments, or student groups. At Tufts, we asked vendors to provide experts to defend their products in a public debate comparing paper and polystyrene cups.

**Make Vendors Aware of the University's Environmental Commitment**

A first step in greening the university's purchasing habits is to make vendors and suppliers aware of the importance the university places on environmentally sensitive products, reduced waste, and recycled goods. Purchasing departments can do this by sending a letter to all vendors that outlines the university's commitment, explains that environmental criteria will be an ongoing consideration in purchasing goods and services

**Box 4.2**
Sample letter to vendors

Dear Vendor:

Green State University is committed to the protection of our natural environment through our operations and in our academic activities. The university has endorsed two environmental stewardship goals: to reduce our generation of solid waste by 30 percent and to hold our carbon dioxide emissions at 1990 levels as agreed in the United Nations Environment Conference in Rio. To this end we are taking steps to reduce and recycle waste, buy local products, and reduce the hazards that we contribute to. In addition, we are very interested in products and the delivery of products that reduce the use of fossil fuels.

We invite your company to join us in a continuing evaluation of products, services, and practices to ensure that we are minimizing our environmental impacts to the extent possible.

Please let us know what steps you have already taken to reduce environmental impacts. We also invite you to meet with our student interns who are well versed in the principles of environmental protection and regularly evaluate new products for us to determine their environmental impacts.

Sincerely,

Susan Green
Director of Purchasing

for the university, and invites them to join with the university in improving the products and services they offer. Box 4.2 shows a sample letter. If the university has embraced broad environmental goals, these goals should be included in the letter. Sometimes this message will be new to the vendor, but purchasing managers may discover that their vendors already have extensive initiatives underway. In addition to improving the enviornmental performance of the products at the university, a letter to vendors encourages those whose environmental consciousness is just beginning and supports those that have already shown a commitment.

Universities may sponsor a meeting with vendors and contractors to explain the university's interest in reducing environmental impacts and the university's environmental policy. This meeting can be used to develop specific action steps for environmental stewardship in dealings with vendors and to discuss recent environmental improvements by both the university and the vendors. Vendors can explain their plans for ensuring that the products they distribute and the services they provide meet or exceed university environmental objectives. In addition, the list of actions generated in the meeting can become the basis for product bidding specifications. The Tufts dining services department found that this type of joint meeting of dining managers, environmental advocates, and food vendors was a productive process for increasing the offering of environmentally friendly food products.

**Add Environmental Criteria to Bidding Specifications**

Universities use a competitive bidding process to obtain the best price for many products and services that they buy. The bidding process usually begins with the purchasing department's generating a written request for bids for a specific product or service. The department evaluates the resulting bid proposals to see that they meet the specifications requested and to compare prices. Because this process is used to do some quality control of the products, universities can also use the bidding process to require vendors to provide product that are greener.

At Rutgers, Kevin Lyons has rewritten many of the specifications for purchasing to include environmental criteria. He works with students, who research the issues associated with products the university buys. He places a great deal of the burden for the change on the vendors, often asking them to identify products, new packaging, or recycling options.[3]

To promote environmental action from vendors, it is often easiest to add criteria to existing bidding specifications rather than starting from scratch. In this way, the bidding specifications contain all of the standard requirements for product performance to ensure that environmentally friendly products and services perform as well as the alternatives they replace. To promote environmental stewardship, the bidding specifications should address the following issues:

- *Focus on product performance rather than specific product characteristics.* Most people want paper to write on, dependable computers, and quality copiers. Purchasing departments that focus on providing materials that get the job done will have more latitude to buy for waste reduction and efficiency. Pilot tests can be essential for evaluating the effectiveness of alternative products and services.

• *Remove barriers to environmental action.* Bidding specifications may contain language that prohibits or discourages the use of environmentally friendly products. Words such as "virgin material only" or "new" eliminate some recycled products from consideration at all. For example, specifying bright-white paper color or requiring uniform paper characteristics for all paper sizes (including legal and ledger paper) may discourage the use of or raise the price of recycled paper.

• *List specific environmental issues related to that commodity.* Bidding specifications should include all the environmental criteria related to that product. If the university wants 100 percent recycled paper with 50 percent postconsumer waste, it should say so. Performance specifications that detail the way the product is to perform should also be included; for example, the purchasing department might add language to the bidding specifications requiring that the copy machines, leased from an outside supplier, use recycled paper successfully. Including this provision in the specifications will reduce the chance that copier repair technicians will blame copier malfunctions on the use of recycled paper.

• *Ask for two alternatives.* Bidding specifications can request a dual bid on two alternatives, such as virgin and recycled motor oil. By receiving quotations for two products, the purchasing department can compare the environmentally sensitive products with the traditional ones. Sometimes the results are surprising to those who think that recycled is always more expensive. Some universities may be willing to offer a premium price for environmental products. For example, in order to support recycling markets, Massachusetts allows a price preference for the purchase of environmentally friendly products that perform as well as the products they replace, although the state usually finds it can offer these products at competitative prices without the premium. Receiving a dual bid can help determine if the criteria for a preferential price are met.

• *Include specifications for packaging and waste.* Bidding specifications can specify the recycled content of products or product packaging. These specifications may also request that vendors take packing or used material back to their warehouses for reuse, reclamation, or recycling. For example, computer companies are now designing their products to facilitate recycling and reuse after the machine's useful life. The University of Vermont has a "recapture program" that requires vendors of some products to take back spent product or packaging. For example, the university returns window cleaner jugs to the supplier for refilling, and it reuses plastic garbage can liners as many times as possible and returns them (free of food- or medical-waste contamination) to the manufacturer, which recycles them.

- *Offer long-term contract relationships to promote environmental products.* To ensure markets and cost-effective availability of particular products, the university may want to offer longer-term contract options to vendors of environmentally sensitive products the university is sure to use on an ongoing basis. For example, a long-term arrangement with a single printer can ensure a reasonable price for high-quality recycled paper by allowing the printer to buy larger quantities of the paper and offer it to the university at a lower price than if the paper were bought in ordinary quantities.
- *Ask for life-cycle pricing.* Bidding specifications can ask that proposals for products include the cost of running the equipment or disposing of it after it is broken or obsolete. In this way, energy-efficient computers and other efficient equipment easily become the low-cost products.

### Use Contracts to Reduce the Environmental Footprint

Service contracts provide important opportunities for environmental actions without burdening or trying to change the practices of existing university staff—for example:

- Moving companies provide reusable packing crates instead of disposable boxes.
- Construction contractors remove and recycle their construction debris, boxes for new lights, and other solid waste.
- Delivery trucks limit on-campus engine idling time to one minute.
- Food vendors offer organic and local foods.
- Dining services contractors use reusable serviceware and recycled empty containers.
- Grounds and dining contractors compost yard and food waste.
- Pesticide applicators use integrated pest management techniques.

## Solid Waste Reduction

Solid waste issues are in the news daily and are important because of both the disposal crisis and the energy and resources used in the production of the goods that quickly become trash. Purchasing departments can make policies and decisions that can reduce the generation of solid waste through source reduction, increasing the amount of remaining material that is recyclable, and developing systems for product reuse.

**Select Products Designed or Packaged to Reduce Waste**

Since the major impacts that result from excess packaging of a product occur during its manufacture, and disposal, less is usually better. Reducing the amount of material, particularly its weight, generally provides environmental benefit from cradle to grave.[4] Purchasing departments can include waste as a criterion in the bidding process and make a conscious effort to consider waste generation in decisions to select products. For example, when similar products are packaged differently, preference can be given to the product with less packaging. And when one item is reusable, recyclable, or less wasteful at the end of its life, it can be selected over a similar product.

Sometimes waste-related decisions motivate a change in the way that the university obtains a product. For example, cleaning supplies can be delivered in concentrated form. Other times purchasing for source reduction means that one supplier is chosen over another because it is willing to provide a service, such as the reuse of packing crates or the recycling of packaging containers for a variety of products. If the purchasing department selects products based on lowest cost alone, it should be sure to factor in the total cost of the product by including the disposal of the item or its packaging. Following are examples of purchasing decisions that can reduce waste:

- New furniture shipped in reusable blankets rather than boxes
- Copiers and printers with double-siding capacity
- Bulk goods
- Large-volume dispensers of toilet tissue and paper towels (saves money for labor as well)
- Concentrated cleaners mixed on-site with water from automated mixing stations
- Elimination of secondary packaging, such as shrink-wrap around boxes
- Reusable boxes or crates for transporting products
- Products or packages that departments will reuse for storage or shipping

**Purchase Recyclable Products and Packaging**

Almost every product is recyclable somewhere, but not all materials are recyclable on every campus. To support on-campus recycling efforts and reduce waste, purchasing managers can make choices about the materials, particularly packaging, they select and give preference to materials that are recyclable on the campus. For example, purchasing departments at universities with cardboard but not plastic recycling programs can request that vendors pack office supplies in cardboard boxes rather than in heavy plastic wrap, since the cardboard is recyclable on campus and the plastic is not. On campuses without an established cardboard recycling program that incinerates its trash for energy production, the plastic may be the better choice since it probably has greater energy as a combustible material than cardboard used for the same purpose.

**Develop Systems to Reuse Materials**

One man's trash is another's treasure, and university departments often have extra or used supplies and equipment that are no longer useful to them. For example, a department may have boxes of once-used manila folders but no room to store them, or another department may have mistakenly ordered a printout binder of the wrong size. These supplies can be useful to others on campus if there is a way to exchange them.

The exchange of large quantities of goods and equipment on campus can be cost-effective and relatively unbureaucratic. The University of Arizona and the University of Oregon have dedicated space on campus for "free stores" that take in merchandise no longer useful in its department of origin and make it available to the university community. The University of Oregon controls access to the store by making a key to the store available in a nearby department. In eight months, it gave away about $11,000 worth of office supplies. The recycling coordinator's student staff keeps the space organized and finds homes for large quantities of materials. Student interns may be helpful for making a workable system at any college.

The University of Wisconsin developed the SWAP (Solid Waste Alternatives Program) in about 8,000 square feet of warehouse space. Surplus materials from the university and local state agencies—computers, lab equipment, office furniture, office supplies, home furnishings, and building materials—are sold at the shop.[5]

## Recycled Products

Recycled products are generally kinder to the environment than their "virgin" counterparts. In addition, they usually require less energy and generate less waste in their manufacture. Buying recycled products supports the use of these products and the markets for the recyclables collected on a campus and beyond.

Recycled paper is the most common of the recycled products on the market, but there are many other products as well: construction products such as plastic lumber and some masonry products, automobile products such as motor oil and retreaded tires, and custodial supplies such as rags and trash bags. Table 4.1 lists some commonly available recycled products.

For some products, using recycled materials is the cheapest alternative. Refilled laser printer cartridges, recycled plastic garbage bags, and newsprint for the school newspaper usually cost less than the virgin products they replace. Since cost containment is nearly always a goal of universities and colleges, identifying recycled products that are lower or of similar cost can facilitate the use of these products.

Under the Resource Conservation and Recovery Act (RCRA) and Executive Order 12873, signed by President Clinton in 1993, the federal government has established standards for recycled products in its own purchasing. The new regulations specify standards for twenty-four recycled-content products in seven product categories. These regulations apply to government agencies and government contractors that purchase more than $10,000 of a product. These guidelines, summarized in box 4.3, are applicable to universities wishing to promote the use of recycled products.

### Make It Easy for Departments to Select Recycled Products

If the university is not in the habit of purchasing recycled products, there may be some barriers to introducing them. For example, the simple layout of the form used to order supplies, copies, paper, and other materials can be a large barrier to the use of recycled products. The purchasing department can make sure that these forms show that recycled products are available. Companies that provide office supplies to large institutions

**Table 4.1**
Commonly available recycled products

| Product | Sources | Issues with product purchase and use |
|---|---|---|
| Writing paper products: copy paper, stationery, notebooks, note pads | Widely available; may require special order for specialty papers; cost decreases with volume purchase | Postconsumer waste should be at least 10%; Green Seal requires 25% postconsumer waste for certification<br>Recycled paper may have higher dust content that can affect copy machines; tune machines to accept recycled paper, and use uniform source for recycled paper<br>Some papers have reduced brightness that affect use for some jobs, but there are 50% postconsumer waste papers that have brightness suited for any job |
| Paper towels, napkins, and toilet tissue | Available cost-effectively from manufacturers and distributors; 100% recycled with 20 to 50% postconsumer waste and unbleached papers are available | Often cheaper than virgin and bleached products |
| Building materials: plastic lumber, concrete | Increasingly available | Try on pilot scale. May last longer than conventional products. |
| Retreaded tires | | Buying long-lasting tires (50,000 miles) and maintaining air pressure can reduce the number of used tires |
| Re-refined oil | | Usually contains at least 10% postconsumer oil and 25% re-refined oil base |
| Laser printer toner cartridges | Available from many suppliers; contact the International Cartridge Recycling Association in Washington, DC, for a list | Some recycled toners do not perform as well as others |

**Box 4.3**
Federal standards for the purchase of recycled products

Components of a procurement program (40 CFR Part 247)
1. A recovered materials preference program
2. An agency promotion program
3. A program for requiring vendors to reasonably estimate, certify, and verify the recovered materials content of their product
4. A program to monitor and annually review the effectiveness of the affirmative procurement program.

Designated items (sections 247.10–247.17)
1. Paper and paper products
    a. Uncoated printing and writing paper (section 504 of Executive Order 12873): minimum of 20 percent postconsumer content beginning December 31, 1994.
2. Vehicular products
    a. lubricating oil containing re-refined oil
    b. tires (excluding airplane tires)
    c. reclaimed engine coolants
3. Construction products
    a. building insulation
    b. structural fiberboard
    c. cement and concrete
    d. carpet made of polyester fiber
    e. floor tiles
4. Transportation products
    a. traffic barricades
    b. traffic cones
5. Park and recreation products
    a. playground surfaces
    b. running tracks
6. Landscaping products
    a. hydraulic mulch products
    b. compost from yard trimmings, leaves and grass
7. Nonpaper office products
    a. office recycling containers and waste receptacles
    b. plastic desktop accessories
    c. toner cartridges
    d. binders
    e. plastic trash bags

often can include a special notation or a special section in their office supply catalogs (or catalogs provided by outside vendors) listing recycled products. Universities can request that office suppliers include this designation as a condition of winning the contract for the university's business.

The best way to make the selection of recycled products easy is to make them the default option—the choice that is ordinarily ordered and delivered. For example, when an academic department orders copy paper, it can receive recycled paper automatically (in contrast to requiring a special designation to order recycled products). Making these products the norm is a bold environmental action, but it is the clearest message and, in fact, the easiest to implement. Some purchasing departments can make this choice on their own; others need administrative action for some changes. Students or others advocating for change should ask, "How was the current paper selected as the default?" and follow the same process to advocate for change.

**Specify Postconsumer Content**

Recycled material includes both postconsumer material and other secondary material and preconsumer material. Postconsumer waste is the material that end users have used that has been captured for recycling, such as white office paper separated at desk-side. In contrast, preconsumer material is secondary material or scrap from the manufacturing process, such as scraps resulting from cutting envelopes. Manufacturers have always used this excess preconsumer material in their products so preconsumer waste content is not as important as postconsumer for promoting recycling by the public. When possible, the university should specify the recycled content that is acceptable to it, including pre- and postconsumer minimums.

Most recycled paper available for copy machines contains at least 20 percent postconsumer waste and 30 to 40 percent preconsumer waste. Papers with 25 percent postconsumer waste are available for this use, and some papers can have up to 100 percent postconsumer waste. Many fine recycled writing and printing papers can meet the needs of printers and graphic artists, although they can be costly because the demand for them is low.

The people who service and repair office equipment, particularly copy machines, often blame the dust from recycled paper for the breakdown of their machines. Before accepting that excuse and switching back to virgin paper, departments should remember that copy machines have always jammed and broken (especially in times of critical need) despite the type of paper used. To avoid this excuse, purchasing managers can include a provision for the use of recycled paper when negotiating the contract for copiers or their maintenance.

**Offer Recycled Products in the Bookstore**

The campus bookstore should offer recycled paper of all sizes (notebooks, index cards, and other supplies) at competitive prices to ensure that it is available to customers. Many university bookstores have allowed student environmental groups to post signs promoting the benefits of buying recycled paper products to promote their use. Offering a range of these products sends an important message in support of recycling and its place on campus.

## Purchasing for Energy Efficiency

Transportation and energy use are major contributors to air pollution problems and global warming. University purchasing departments can minimize the energy, transportation, and air pollution effects of their purchasing through the use of local and efficient products, thereby reducing the institution's contribution to these problems. In particular, the purchase of office equipment such as computers, printers, and copiers holds many opportunities for efficiency. Locally manufactured products, fuel-efficient vehicles, and alternatively fueled vehicles can also reduce a university's impacts.

**Buy the Most Efficient Computer Equipment Possible**

The age of technology has resulted in an increased number of office machines that use electricity. The number and use of computer peripherals, such as printers, scanners, and plotters, is also on the rise. With each new computer or printer or each additional hour of a computer's use comes a proportional increase in electricity and its associated costs, including pollution. In fact, the electricity consumption of office equipment is growing faster than any other component of energy consumption in commercial buildings. Office equipment accounts for approximately 5 percent of the total commercial energy consumption in the United States.[6]

Computers and office equipment can be selected and operated to minimize this energy demand throughout the life of the equipment. Determining the full cost of these purchasing and operating decisions by weighing purchase price and energy used is complicated and requires attention to the latest technology. Purchasing departments can specify that departments purchase efficient computers, but they must also work with computer users so that the equipment is operated efficiently (see chapter 6). Student research projects can determine the environmental and financial trade-offs of each of these purchasing decisions.

The full cost of the office equipment includes the cost of running the equipment and also the purchase price. We are accustomed to thinking this way when we include gas mileage in a decision about which car to buy, but we rarely ask how many kilowatts a computer uses, although electricity has both environmental and financial costs that may be dramatic over the life of the equipment. EPA's Energy Star program has set energy-efficiency criteria and awarded the Energy Star seal to computers, printers, monitors, fax machines, and copy machines that meet or exceed explicit efficiency standards by having a "sleep" or low power mode when the machine is idle. Box 4.4 shows the potential savings from using Energy Star computers.

Purchasing equipment with the Energy Star designation is one way to obtain efficient equipment. In the first year of the program, the U.S. gov-

**Box 4.4**
Potential annual savings from Energy Star systems and turning computers off at night

System includes the computer, a monitor, and a laser printer, with costs based on $0.08 per kilowatt-hour.

Conventional system (150 W) always on: $150
Energy Star system (67 W) always on: $47
Conventional system off at night: $25
Energy Star system off at night: $11

ernment purchased over 292,000 computers, 164,347 monitors, and 64,773 printers that met these guidelines; the purchases are predicted to reduce utility bills by an estimated $5 million per year.[7] A single university cannot realize such large savings, but specifying that equipment carry the Energy Star designation can save it thousands of dollars in electricity costs annually.

The amount of electricity computers and their monitors use is generally proportional to their size, speed, and features. For example, an 80486 machine generally uses more electricity than a 80386, and a monochrome monitor uses less than a color monitor. Whenever possible, those buying computer equipment can obtain the data on the actual electricity use. Upgrading existing equipment rather than buying new is usually more efficient, cost-effective, and generates less waste. When buying new, laptop and notebook computers usually use less electricity than desktop models. The EPA Energy Star Compliant Product Database lists office equipment that has met the Energy Star criteria. Participating manufacturers in the program self-test equipment and voluntarily submit compliant products to EPA. The product list provides equipment specifications and features and is updated monthly. The home page lists provide the complete list on-line.[8]

Computer printers also use vastly different amounts of electricity. In general, dot matrix printers are the most efficient printers, when printing and when idle. Certainly student papers and drafts of faculty research, draft memos, and accounting records could easily be printed on dot matrix printers, but we have all become increasingly accustomed to the speed and polished look of higher-quality printers. Laser printers are energy intensive, but printers with similar features can have quite different energy demands. Ink jet printers provide a relatively energy-efficient technology with a quality and speed similar to that provided by laser printers. The ink jet printers typically meet the Energy Star criteria, largely because they use much less electricity when idle. Ink jet printers are also less expensive than laser printers, and the technology is always improving.[9] Box 4.5 shows how to select computer equipment for maximum efficiency.

**Buy or Lease the Most Efficient Copiers Possible**

Copiers are generally the largest pieces of energy-using office equipment in a university academic or administrative department, and a single ma-

**Box 4.5**
Maximizing efficiency in computer equipment

Computer
- Buy a lap-top computer.
- Buy an Energy Star computer.
- Upgrade equipment rather than buy new.

Monitor
- Buy an Energy Star monitor.
- Buy a monochrome monitor.
- Buy monitors only as large as needed.
- Buy only as much resolution as needed.

Printer
- Buy an Energy Star printer.
- Consider an ink jet printer.
- Share printers when possible.
- Consider a double-sided printer.

*Source:* Based on Marc Ledbetter and Loretta A. Smith, *Guide to Energy Efficient Office Equipment,* EPRI TR-102545 (Berkeley, CA: American Council for an Energy-Efficient Economy, 1993).

chine may each consume nearly 1,600 kWh annually.[10] Many universities standardize copiers, installing a single brand and model throughout the campus. Thus, saving even a few watts on each machine can result in significant savings. Purchasing managers can obtain energy ratings from copier manufacturers in order to compare machines.

Correctly sizing a copier to meet the needs of the office without being oversized is particularly important since energy use is greater with larger copiers. When possible, the university should select copiers with power management capabilities that reduce the machine's demand for electricity during periods of inactivity. Copiers with the Energy Star designation are more efficient models and have a "sleep" feature that reduces their electricity consumption when they are not in use.

Copiers and printers that can copy or print on both sides of the page can save energy and reduce waste, although they use more energy than their single-sided counterparts. (The energy that is used to make the paper—its "embodied energy"—is greater than the incremental energy used to make a double-sided copy.)[11]

### Turn on the Power Management Feature on Equipment

The fact that office equipment is equipped with energy-saving power management features is no assurance that it is being used. University policy should dictate that these features are enabled and that university computer and copier repair staff are trained on this feature.

### Include Energy and Water Efficiency as Criteria for the Purchase of Equipment

Universities and colleges purchase, lease, and use all manner of equipment, from microwave ovens to laundry machines, vending machines to bottled water dispensers. Minirefrigerators in small offices are notoriously inefficient and run constantly. Each of these many pieces of equipment uses electricity and sometimes water. Paying close attention to the efficiency of this type of equipment throughout the campus can have significant savings. In some cases, selecting the most efficient equipment may require paying more initially for the machines, realizing savings in the equipment's operating expenses. Sometimes the change will require that the purchasing department work closely with manufacturers to modify efficient equipment to meet the needs of the university. For example, Tufts found that the most energy-efficient laundry machines available were not equipped with card readers for students' prepaid debit cards, but the manufacturer was willing to make the modification so that the more efficient machines could be used in ways that students desired. The green university can consider efficiency among its most important criteria when purchasing energy-using equipment.

### Give Preference to Local Products

Buying goods and services from local farmers, manufacturers, and contractors can reduce the number of miles traveled to and from campus by trucks and other vehicles. Since vehicle emissions are a major cause of air pollution, reducing the length of delivery trips can improve air quality on the whole. Buying from local vendors supports a university's regional economy, a choice with political benefits as well. Furthermore, university business can be a sufficient boon to small and local companies that they may offer discounts or be able to eliminate distributors, brokers, or other middlemen. For example, buying meat products from a local farmer can help secure his markets and provide him with an outlet for some of his less popular cuts of meat (e.g., pork backs used in soups).

## Printing Services and Copy Centers

Many colleges and universities have on-site printing and copying centers that handle the vast majority of the university's needs. Often these centers are managed by the purchasing department or operate in the same administrative unit. The centralized nature of these facilities offers a range of opportunities to respond to environmental issues.

### Make Environmental Choices the Default

Printing and copying service centers on campus see themselves as service providers, there to offer a specific service on request. But these services also have many opportunities to take active roles in reducing waste and promoting recycled paper.

Most printing centers provide single-sided copies on virgin, white copy paper on an ordinary (or default) basis. They can change this default to double-sided, unbleached, recycled paper. If customers want something other than this default, they must specify their need in the same way that those who now want recycled paper often must specify it. Most customers do not care what is used if the cost differential is negligible and the quality meets their needs. The center may want to tell customers about the change, giving the reasons and motivation for it. Or it may want to wait and see if anyone even notices; Tufts made this change, told no one, and received no complaints.

Making double-sided copies on unbleached, recycled paper on a regular basis should be encouraged (with price if possible) and advertised on the order forms and at the drop-off counter, since many people are not aware of these options. Students, recycling coordinators, or members of the environmental committee may take on the training of the printing and copying staff so that they understand the issues and trade-offs.

### Reduce, Reuse, and Recycle

University printing and copying facilities generate large volumes of waste paper, since copying and printing machines seem to fail often. Printing departments can reduce waste by reusing scrap paper in test runs, making notepads from scrap paper (called *precycle pads* at Tufts), and providing shredded scrap paper to other departments or to off-campus companies that use it as packing material. Kurt Teichert, environmental ombudsman at Brown University, even has business cards printed on reused scrap paper.

For recycling to be successful in print and copy shops, there need to be effective collection facilities for recycled paper as well as a reliable schedule of pickups for the collected paper. In addition, employees must be trained on the types of paper that are recycled, and periodic checks should be made to ensure that the system is effective. Coordination with the buildings and grounds department is important.

**Reduce Exposure to Toxics in Printing Facilities**

The large-volume printing and copying machines found in university printing facilities can make working conditions unpleasant and even unhealthy. EPA has proposed regulations that would regulate air quality in large printing establishments. Although smaller shops, such as those found at most universities, would be exempt, these shops can benefit from improving air quality. Universities must ensure that these facilities are well ventilated and that toxic emissions are removed from the workplace.[12]

The inks traditionally used in printing and copying machines contain volatile organic compounds that are hazardous to worker health and contribute to air quality problems. Many of these printing inks have a petrochemical base and produce vapors that can be harmful to workers. Changing to inks with a vegetable base, usually soybean oil or other vegetable oil, reduces the exposure to volatile organic compounds and the dependence on petroleum.

As with recycled papers, there are few standards for soybean or other vegetable oil inks. Some reduce the amount of petrol-based solvent only slightly, and it is rare that petroleum products are completely replaced.[13] Designations such as "Printed with Soy Ink" simply mean that some soy oil was used. The National Association of Printing Ink Manufacturers recommends that vegetable oils be a minimum of 20 percent of total ink ingredients.[14] The American Soybean Association certifies inks by using the "SoySeal" trademark.

Solvents are commonly used to clean inked rollers between printing jobs. Alternatives to these solvents include soap and water or solvent-free solutions. Dedicating certain presses to specific inks or groups of ink colors can also decrease the need for solvents by reducing the frequency with which rollers need to be cleaned.

**Reduce Paper Use in Printing Activities**

In some universities, nearly every academic department, school, and student group has its own newsletter or magazine. Some publications target prospective students, while others reach out to alumni and staff. There are many opportunities to reduce waste from these university publications and to reduce printing and mailing costs as well. In fact, electronic mail may eventually replace many printed newsletters.

The communications office of the university can ensure that each person on the mailing list receives only one copy of a given publication. Although this seems self-evident, sometimes an alumna working for the university receives three copies of the university newspaper or the annual message from the president: one as a staff member, one as an undergraduate alumna, and a third as a graduate alumna. If she is married to an alumnus or faculty member, even more duplicate copies may come to the home. Computer science students may be able to help out by writing programs to coordinate the university mailing lists.

Accurately estimating the number of copies of each publication to be printed can also reduce waste and costs. Slightly underestimating the number of publications needed on a dozen jobs will make up for the cost to reprint a single job where the shortfall is important. Print overruns or free copies provided by some print houses because the paper stock is already in the press should be avoided when possible.

Student newspapers are notoriously overprinted. Student governments that fund these papers and other student groups could take action to reduce waste by making a "print what you need" policy as a condition of funding student newspapers.

## Vehicle Efficiency and Alternative Fuels

Universities' own gasoline- or diesel-powered cars, trucks, vans, plows, and bulldozers emit air pollutants—hydrocarbons, carbon monoxide, nitrogen oxide, and sulfur dioxide—contributing to the air quality problems in the region. For example, in Massachusetts, mobile sources, such as cars and trucks, generate 48 percent of the state's volatile organic compounds, 47 percent of its nitrogen oxides, and 70 percent of the carbon monoxide. In addition, almost one-third of the state's carbon dioxide, the principal human-produced greenhouse gas, comes from cars and trucks.[15] The effect is similar throughout the country.

The selection of vehicles is an important opportunity for a university to reduce the amount of fossil fuels burned (saving money) as well as providing a demonstration site for new technologies. The purchasing department can develop a vehicle purchasing policy that promotes efficiency and the use of new technologies. The department can screen all vehicle purchases for conformance to the policy.

**Buy the Most Fuel-Efficient Vehicles Possible**

Selecting vehicles to maximize fuel efficiency is the most obvious way to reduce the university's use of fuels. Smaller vehicles, vehicles with manual transmission, and those that do not have air-conditioning tend to be most efficient. Tufts found that it could dramatically improve efficiency and reduce cost by purchasing small trucks rather than full-size trucks (half ton and larger) for most of the needs of the grounds departments.

**Select Alternative Fuels When Possible**

Vehicles powered by alternative fuels provide opportunities to reduce the university's impacts on air quality and climate change and support the development of new technology. Furthermore, state regulations to implement the Clean Air Act (State Implementation Plans) may mandate that centrally fueled vehicle fleets, such as a university's, use "clean" fuels in some portion of their fleets. Fines for noncompliance may eventually reach several thousand dollars for each occurrence. An educational institution, with a stated commitment to environmental protection, is ideal for an alternative-fuel demonstration program that shows the effectiveness of these new technologies, provides student research opportunities, and helps to increase public acceptance of these technologies.

There are several alternative fuels suitable for powering vehicles: electricity, compressed natural gas (CNG), reformulated gasoline, oxygenated additives, and other fuels, including hydrogen, fuel cells, ethanol, methanol, and propane. The feasibility of switching from gasoline or diesel to one of these fuels depends on several factors: (1) the ability of the vehicle to meet university needs, (2) fuel availability, (3) cost, (4) maintenance considerations, and (5) emission from the vehicle.

In the near term, CNG and electricity are probably the most viable fuels for universities, although reformulated gasoline is increasingly common in areas with air pollution problems. Harvard University plans to use an electric vehicle to collect recyclables on the central quad of the campus. The short trips of the recycling route and the need to reduce the smell of exhaust in pedestrian areas make this an ideal application for an electric truck. The University of Southern Florida has a solar charging station for its fleet of four electric trucks.[16]

State offices of transportation can identify programs to pilot-test these alternatively fueled vehicles. Local natural gas companies or electric utilities can also help universities to find vehicles that are suitable for the terrain and use available fuels. Since the use of these low-emissions vehicles is experimental and unfamiliar, universities should try to see if the utilities or the state will help underwrite and publicize a project to implement alternatively fueled vehicles on campus.

## Hazardous Materials in Laboratories and Studios

Universities use a variety of toxic chemicals and materials that contain harmful ingredients. The departments that use these substances on campus include chemistry, engineering, biology, physics, geology, art and photography, and buildings and grounds. Schools of medicine and dentistry also purchase and generate large quantities of hazardous materials. The purchasing department can be a valuable gatekeeper for encouraging the purchase of less-toxic alternatives, ensuring that proper management occurs and facilitating chemical reduction and exchange. Many of the techniques already described in this chapter—using suppliers and the bidding process, mandating certain changes, and making change easy—are effective in dealing with the purchase of chemicals. Efforts in this area must also be coordinated with the individual or department responsible for health and safety, as well as with the faculty in departments that use chemicals, in order to ensure that regulations and oversight are in place.

Individual (and often tenured) faculty members, who have a great deal of autonomy, are often resistant to change and reluctant to relinquish purchasing decision making on chemicals. The culture of most faculty and academic departments is uncomfortable with direction or mandates from administrative departments such as the purchasing department or the environmental health and safety department.[17] If lower overall costs are a motivation, purchasing and environmental health and safety departments can team up together to make a powerful case for reducing chemicals on campus. In turn, this effort will reduce hazardous wastes and accidents—costly for any university.

**Use the Services of Chemical Suppliers to Reduce Hazards**

University laboratories are large purchasers of laboratory chemicals for use in teaching and research. Ordinarily most of these chemical purchases are ordered through the university purchasing department and from one of several vendors on contract with the university. However, faculty and departments that use chemicals or other hazardous materials often have an open purchase order with a chemical supply company that allows them to buy whatever they need, with few checks and balances.

One problem for the university is that chemicals are dramatically cheaper when bought in large quantities. Ordinarily bulk purchasing is consistent with environmental objectives, since it reduces packaging and shipping. However, with chemicals, the cost of disposing of their packaging is minute compared to disposing of the chemical itself or the waste products that result from their use. Researchers, technicians, and graduate students, anxious to get a good deal, may buy more product than they need. The result is that additional product is stored in the laboratory and ends up as hazardous waste.

University purchasing departments should work with departments using chemicals and the environmental health and safety staff to change the bidding specifications for selecting chemical suppliers. Suppliers doing business with the university can help in this regard by:

- Providing designated hazardous materials (the environmental health and safety department or manager can provide a list) in break-resistant containers of metal, plastic, or plastic-coated glass. The cost per chemical is slightly higher, but the reduced risk of spill, and its associated financial and health costs, are dramatic.

- Providing an information card packed with designated hazardous materials indicating their recommended use quantities and any special instructions.
- Storing hazardous materials that the university has purchased at the supplier until they are needed at the university. If the materials are not requested by the university within a designated time (such as one year), the material reverts to the chemical supplier, which can resell the material. For example, if a department buys six liters of a chemical, five liters could remain with the supplier and shipped liter by liter as needed. After a designated period of time, the unshipped material becomes the supplier's property and it can be reinventoried, thereby saving the university the costs of disposing of extra chemicals and allowing resale of the hazardous materials. (In fact, implementation of this strategy is most likely to force researchers to buy chemicals in quantities that they actually need.)
- Taking back single-use, difficult-to-handle items, such as one-way non-returnable, compressed gas cylinders, at the end of their useful life.

### Track Chemical Purchases

Purchasing departments can help promote safe chemical handling by tracking all chemicals that the university purchases. The purchasing department helps to track chemical use by recording for each substance its Chemical Abstract Service number, a unique number that identifies each chemical, when it is purchased by individual labs, researchers, or departments. The purchasing department can then provide labs with an annual summary of the chemicals purchased. Maintaining an accurate inventory of chemicals on hand assists in the laboratory safety audits, chemical inventory efforts, and waste disposal programs.

Tracking chemical purchases can be particularly useful for identifying new chemical users on campus. For example, an engineering faculty member may begin new research requiring the use of a strong acid. The purchasing department can help the environmental health and safety department identify that user to ensure that the laboratory facility is appropriate, that wastes are handled properly, and that the professor and his or her staff has the required safety training.

### Balance Solid Waste Reduction and Safety

Because so much of our environmental attention has focused on recycling, and to some extent reducing, solid waste, it is important to realize that there are many times when the trade-off between solid waste generation and safety favors safety. For example, although glass containers may be recyclable, plastic containers are less breakable and can prevent chemical spills. Similarly, efforts to track chemicals and provide information about safety and prudent waste management are valuable for promoting safety even if they generate some additional paper waste. Of course, streamlined systems that provide only the necessary information are easier to use and generate less waste and should be encouraged.

### Enforce Safety Rules as Chemicals Enter the University

Purchasing departments may serve as gatekeepers, identifying potentially dangerous chemicals that are entering a university. Although purchasing managers usually do not have extensive knowledge of chemicals and their properties, the system for purchasing chemicals can help to keep the department or individual responsible for overseeing environmental health and safety staff up to date regarding the location, types, and quantities of hazardous materials. In addition, the purchasing department can help to enforce rules about maximum quantities of materials, proper transport of materials, or laboratories or researchers who may be restricted in their use of particular substances. For example, to improve safety and reduce the incidence of chemical spills, a purchasing department might require that transported chemicals be purchased in break-resistant containers or require that students who purchase more than five milligrams of a chemical meet with a safety officer who evaluates the request. Purchasing might penalize departments that insist on buying chemicals or other materials through undesignated vendors.

### Make Surplus Chemicals Available for Exchange

Sometimes surplus chemicals or chemical by-products from one laboratory or department may be useful to another department or laboratory. Sharing chemicals can save money and reduce waste generation. Where possible, the purchasing department can facilitate this exchange and work with the environmental health and safety department to ensure that the exchange program complies with all necessary regulations and that chemicals are transported safely and legally. Chemical exchange among researchers and faculty in the same department may be undertaken on a slightly less formal basis.

## Nonlaboratory Hazardous Materials

A purchasing department can serve as the catalyst for university action to reduce the use of nonlaboratory hazardous materials, whether they are regulated or not. Many of these materials are commonplace, so it is easy to overlook their health hazards and off-site environmental impacts and undervalue the opportunities for reducing them. Items such as cleaning products and paints may contain chemicals that may cause respiratory irritation to building occupants and particularly to workers who use them, or may harm water bodies in their disposal. The green university will reduce its use of such products.

The actions listed in this section are only a beginning of the many initiatives that the purchasing department may wish to investigate in its efforts to evaluate products with fewer impacts. As new research and products are available, the department can play an important role as an advocate and agent for the environmental health of the university.

### Eliminate the Use of Ozone-Depleting Chemicals

Ozone-depleting chemicals such as chlorofluorocarbons (CFCs), used as propellants in aerosols, halon used in fire extinguishers, and refrigerants are regulated and will soon be banned by the Clean Air Act of 1990 and the Montreal Protocol. Despite the bans on production, there are many places where ozone-depleting chemicals are in use on a campus, primarily in the refrigeration and air-conditioning units (see chapter 3). CFCs are also found in academic departments, such as the electrical engineering department, where they are used as cleaning compounds. The University of Kansas determined that ozone-depleting chemicals were also used for preserving documents in the library. The purchasing department can take a leadership role by requiring university departments and contractors to exceed current guidelines and by helping departments to identify alternate technologies.

Purchasing managers reduce the amount of CFCs or other ozone-depleting compounds by asking vendors to notify the university if their product contains these substances. Often alternative products are readily available. State offices of environmental protection can help find these products.

### Reduce the Hazards of Painting

Paint, varnish, stain, paint thinner, and paint stripper are hazardous materials. Paint fumes can cause headaches, stuffiness, and more severe symptoms in people who have chemical sensitivities. Waste paint, paint thinners, and turpentine must be disposed of as hazardous wastes.

As with laboratory chemicals, the purchasing department can help to ensure that paint and paint supplies are managed by using the same strategies that are used for chemicals. For example, when a gallon of turpentine is purchased by the buildings and grounds department, a small "tax" might be assessed to help offset the cost of disposal. The purchasing department might collect the tax and pass it along to the environmental health and safety department. Another alternative is to create a paper trail (or computer trail) that alerts the environmental health and safety department that waste will be generated.

As always, waste reduction is the first step in a paint management program, so buying only the paint that is needed can reduce risk and disposal costs. The purchasing department can help departments order only what they need. Often paint and other supplies are overpurchased because of the high cost of idle labor while waiting for paint deliveries. Purchasing departments can reduce paint use and costs by facilitating rapid reorder and delivery of paints. By working with the buildings and grounds department and contract painters, the purchasing department may keep accurate and up-to-date inventories of materials already on hand.

Working with the buildings and grounds department, purchasing can restrict (or at least monitor) the types of paints that contractors and employees use. For example, latex paints are less toxic than oil-based paints, and some latex paints are nearly odor free.[18] Purchasing departments might also arrange a paint exchange program so that departments, including the art and theater departments, can buy and use paint left over from painting jobs across the university. Leftover paints that are not needed on campus may be useful to charitable organizations, such as Habitat for Humanity, community groups, or local public schools. Some companies collect unused paint for reformulation and resale.

### Reduce and Recycle Batteries

Over 2 billion household batteries are sold annually in this country, resulting in an estimated 83,500 tons of discarded batteries.[19] Batteries are of environmental concern because they contain heavy metals—mercury, lithium, and cadmium—that can leak from corroding batteries, causing irritation to skin and eyes. In addition, batteries corrode and leak in landfills, contributing to the contamination of groundwater. When incinerated, those heavy metals can become airborne.

University purchasing departments can reduce the use of batteries and the environmental impacts associated with their disposal by purchasing rechargeable batteries, collecting used batteries for recycling (perhaps by using a deposit system), asking vendors of battery-operated products (such as pagers) to make products available that can be operated by rechargeable batteries, and/or to take back used batteries for recycling. Developing appropriate collection systems is the most difficult task, but offering an incentive to turn in used batteries when receiving new ones might help.

Rechargeable batteries have limitations; many can be recharged only a finite number of times (sometimes as few as twenty) before they reach the end of their life. Technology is increasing the life of batteries and the utility of rechargeables, and it is also offering alternatives to batteries in solar calculators and other small solar-powered machines. Purchasing departments should take steps to purchase these less harmful products and eliminate their use whenever possible. Since battery collection for recycling is still uncommon in the United States, the truly green university might establish a battery recycling program and open it to batteries from students and staff as well.

**Avoid Sources of Indoor Air Quality Problems**

Many indoor air quality problems result from building construction and air-handling systems (see chapter 3), but there is ample evidence that furnishings, carpeting, and equipment also contribute to these problems. Chemicals, including 4-phenyl-cyclohexane, formaldehyde, styrene, acetone, methylcyclopentane, N-octane, N-undecane, N-dodecane, toluene, xylene, and a number of different benzene compounds can cause eye and nasal irritations, headaches, nausea, and allergic reactions. These chemicals are prevalent in glues and adhesives, backing on carpets, and as fabric treatments in various office and building products, such as carpeting, chairs, draperies, pressed wood counters and desktops, walls and wall panels, shelves, and cabinets. The chemicals cause problems as they "off-gas" or are released to the air. For example, formaldehyde adds permanent-press qualities to fabrics such as draperies and is an ingredient in the adhesives in plywood and fiberboard. Exposure to formaldehyde can cause eye, nose, and throat irritation, respiratory problems, skin irritation, and allergic reaction. High levels of it may trigger asthma attacks, and it is possible for people to develop sensitivity to formaldehyde following sustained exposure.

Purchasing departments can specify that new furnishings be allergy free or otherwise designed to reduce chemical exposure, particularly of formaldehyde. Natural fiber fabrics, furnishings of solid wood or metal, and items that are fastened rather than glued are less likely to cause chemical sensitivity symptoms.

### Find Alternatives to Chlorine

Chlorine can have severe health and environmental effects on human and wildlife populations, especially near manufacturing locations with high chlorine use such as paper mills or plastics plants. Also, the incineration of products containing chlorine, such as PVC plastics, results in dioxin, a carcinogen. When chlorine bleach is used in university laundries or custodial services, the helpful bacteria in the wastewater may be killed. Reducing or eliminating chlorine bleaches and using vinyl (contains vinyl chloride) and number 3 plastics can reduce the university's contribution to this problem. Reducing the use of bleach and bleach products for custodial purposes and for cleanup in laboratories also reduces mercury emissions. Purchasing unbleached writing paper, paper bags, and sanitary paper products also helps reduce chlorine use. Students can help identify products that provide safe alternatives to existing products bleached with chlorine.

## Purchasing as a Source of Environmental Information

The purchasing department has an important educational role to play. Even in a decentralized university, it is often a central point of information, policy, and process for all items bought within the university, and it can therefore be a university-wide catalyst for action to minimize environmental impacts.

### Form a Student-Purchasing Eco-Partnership

Environmental issues are complex, and our understanding of them is constantly changing. New products and new technologies come on the market each week. Purchasing managers who are interested in environmental stewardship need to research the most pressing issues and remain open-minded about the alternatives. At the same time, purchasing managers should remain wary of the product performance needs, costs, and false environmental marketing claims.

Students are a ready resource for identifying alternative products that reduce waste, improve efficiency, or are less hazardous. Working with the purchasing department, students can gather data, verify manufacturer and vendor claims, and help to pilot-test products. In order for a partnership between students and the purchasing department to be successful, students must be given guidance about product performance expectations and cost considerations. In return, purchasing managers should pledge to evaluate and try some of the student recommendations. Box 4.6 shows some examples of possible topics for a student-purchasing eco-partnership project. These partnerships may seem to be farfetched to managers who have not tried them, yet they are working at a number of universities. The keys to their success are clear communication, willingness to try new ideas, and realistic consideration of product performance, availability, and cost.

### Use the Central Supply Catalog to Promote Environmental Choices

Many university purchasing departments publish a catalog of supplies or work with one or more suppliers (most often for office supplies) that provide catalogs of supplies for the university. One way for the purchasing department to serve as a catalyst for environmental stewardship is to use a catalog of supplies that are centrally available as a resource guide for environmentally friendly buying by including a section or specific notation of products that have less packaging, are available from local supplier, or are made of recycled material.

### Identify Alternative Products

When members of the university community request materials that have more environmentally friendly alternatives, purchasing department staff

**Box 4.6**
Sample projects for a student-purchasing partnership

1. Investigate specific recycled products.
   Suppliers
   Price
   Price for quantities used at the university
   Pilot test
2. Investigate vendor contract specifications.
   Research how other universities have included environmental considerations.
   Draft language to be included.
3. Draft memo or e-mail to university community with information on environmentally sound products.
4. Identify problems with existing environmentally friendly products.
   Verify problems.
   Conduct tests (e.g., of copier jamming rates for recycled paper).
   Develop specific solutions.
5. Research how government offices in the area are implementing the federal mandates to buy recycled products. Determine how these strategies can transfer to the university.
6. Promote environmentally friendly products.
   Write stories for the newspaper.
   Publicize ways that purchasing changes can save money.
7. Combine purchasing education with waste reduction education.

can point out alternatives and provide reasons for their use. The purchasing department may suggest alternative products such as recycled paper or a product that is packaged to reduce waste. This is not hard for products that the university buys in large quantities and distributes to departments in smaller quantities, such as copy paper or cleaning supplies. Products bought directly from a supplier and delivered to a department are more difficult to regulate. If the university has a program to exchange used or surplus supplies or equipment among departments, the purchasing department may be able to evaluate purchasing requests for new products against the listing of surplus on hand.

At MIT, the Safety office and the office of Environmental Medical Services asked the purchasing department to help educate faculty and staff researchers using hazardous materials about source reduction (using smaller quantities of chemicals) and techniques to reduce waste and accidents. To do this, they published articles in appropriate university publications (e.g., *Tech Talk*), held symposia on waste reduction, focused on management solutions (e.g., regular inspections and clear policies), distributed a source reduction publication such as *Less Is Better,* and made "Reduce, Reuse, Recycle" a theme for purchasing.[20]

**Designate an Environmental Expert**

It would be ideal to have all members of the purchasing staff knowledgeable in the environmental issues at the university, but it is more realistic to designate a single member of the department staff who can remain informed about new products and monitor the environmental issues associated with purchasing decisions. In this way, at least one person can answer questions and keep an eye out for new opportunities. Most likely, this person will be an existing staff member who has an interest in the issue. At very large institutions, it may be possible to hire an environmental specialist who is charged with helping the department to meet state, local, or university-wide goals. For example, the Massachusetts Department of Procurement and General Services employs an environmental specialist to evaluate purchases and contracts throughout the state as part of the governor's Green State Initiative.

**Publicize Changes to Environmentally Friendly Products**

Changing to environmentally friendly products, like other environmental efforts, is worthy of notice. Purchasing departments can publicize these changes inside and outside of the university. This publicity is a way to encourage those who have taken time to make changes, change the beliefs of others, and remind everyone of the university's commitment and responsibility to environmental protection in conducting its daily business.

**Establish University Environmental Committee for Purchasing Decisions**

A university-wide environmental committee can be helpful in supporting and promoting stewardship in university purchasing. Because the purchasing department strives to be customer oriented, it often has difficulty mandating products that should be used (such as recycled products or products with low toxicity) or developing policies that not everyone may accept. The committee can be helpful in supporting or institutionalizing these changes. In addition, the committee will find that many of the actions that it wishes to undertake require a change in the way that a good or service is purchased. Representation from the purchasing department to the environmental committee will help to ensure that the necessary purchasing changes are made.

## Conclusions

College and university purchasing departments can make decisions and adopt policies that reduce waste and improve the use of resources. Their role can include offering alternative products, advocating for their use, and mandating the use of certain products. Working with vendors to change the products offered on one campus can help other universities and institutions that buy from the same vendor. While environmental advocacy seems contrary to the purchasing culture in most universities, there are countless examples where mandated products and procedures have little or no basis. Certainly purchasing departments need to be careful and reserve their power of mandate for issues that are sure to succeed (e.g., recycled paper) or will have dramatic cost savings or liability reductions for the university as a whole. Nonetheless, the purchasing department can join all the others in taking a leadership role to reduce the environmental footprint of the university.

# 5

# Dining Services

Many resident college and university students visit campus dining rooms and cafeterias more frequently than they enter classrooms. In addition to the meals served to students, university dining services provides thousands of meals to faculty, staff, nonresident students, and visitors. Some colleges and universities operate and manage their own food service departments, but contracting with outside food service providers is increasingly common. Regardless of who is responsible for feeding the university community, the process of growing, preparing, cooking, and serving food has environmental consequences. Solid waste results from packaging and food. Cooking and storing food uses energy from gas and electricity. Pesticides and fertilizers in crop production and the use of chemicals in food processing have an impact on health, ground water, and soil quality.

There are many opportunities to green university dining services. In fact, the dining services operation holds great potential for environmental action because the department as a whole (the management of all kitchens and dining halls) or in individual dining units (such as a kitchen, hall, cafeteria, or canteen) usually has control over decisions with environmental impacts, from meal selection and food purchasing to waste disposal. Furthermore, more than any other department on campus, dining services is used to meeting deadlines and delivering a service—three times a day. This culture makes dining an opportune and rewarding place to undertake environmental stewardship initiatives. Most of these actions, however, require educating the consumers, vendors, dining hall staff, and management.

The primary objective of university food services is to provide quality meals to the university community and its guests. As a result, the quality and variety of food, the quality of the service, and the cost-effectiveness of the program determine the department's performance. Logically, dining staff, managers, and their food vendors are often slow to change and reluctant to try new ideas if there is any danger of impeding the delivery of meals. Efforts to reduce waste, conserve energy, eliminate pesticides, and minimize transportation miles per meal must complement the traditional measures of successful food service operations: quality, cost, delivery, and customer satisfaction.

Dining services' customers' needs, habits, and expectations are often barriers to making some environmentally sound decisions. Student and staff demand for take-out foods have increased the use of disposable dishes. The demand for fresh vegetables and tropical fruits in salad bars and on vegetarian menus throughout the year requires that foods be transported from great distances. In addition to the ecological cost of transporting these foods, such perishables are often treated with preservatives or maybe even genetically altered to ensure freshness and taste at their final destinations.

Across the country, college and university dining services departments are finding that improving the environmental impacts of their operations can increase business and boost employee morale. At Tufts, we found the dining services management and staff among the easiest to work with and the most receptive to our suggestions, largely because of the leadership of the dining director and the service-oriented nature of the dining profession. Dining managers were accustomed to hearing and addressing students' concerns, and their environmental interest initially grew from student interest in reducing the use of disposable cups in eat-in and take-out dining halls.

## Finding Environmental Stewardship Opportunities

Because university food service operations are relatively confined to kitchens and dining rooms, they lend themselves well to initial and ongoing audits to identify areas for environmental improvement. As I discussed in chapter 2, these audits must be done purposefully. An initial audit should be undertaken in order to familiarize members of the audit team and others who are interested (such as students or an environmental advocate) with how the dining units function and to identify initial opportunities for action. The initial assessment can also be helpful for educating the dining managers and staff about areas where their operations have environmental impacts. This initial walk-through is particularly important for finding and celebrating areas where environmental action is already well underway. Many more detailed assessments will be needed to tackle any facet of environmental action completely.

### Conduct a Walk-Through Assessment

An assessment or audit can be done by an individual or by a team, but it will be most productive with a small team that includes members with knowledge of environmental issues, dining decision makers (a unit manager or dining director), and someone with day-to-day knowledge of operations (a manager or a kitchen staff member). A walk-through of a kitchen and loading dock area can take several hours.

**The Loading Dock** Dining services receives food and supplies and stores wastes and recyclables on the loading dock. Most of the products that enter the kitchens are in cardboard boxes or cases, and plastic. An assessment team can peer into the trash dumpster and recycling bins to confirm that these materials are present along with a few others. The auditors can also assess the effectiveness of any existing recycling program by looking in the containers and the dumpster for each material. If the university does not have a recycling program, looking into the dumpster will also confirm that a large portion of the trash is probably corrugated cardboard. The team may identify the ways that the loading dock is currently used and the challenges faced daily, such as space, traffic, or vandalism. Sensitivity to these realities will help ensure that new environmental action programs will work.

Loading areas usually provide access to a wide doorway. Has dining services installed or tried using air curtains to reduce the heat loss or gain from outside sources? Are the lights in this area on unnecessarily? Could more efficient lighting be installed?

**Food and Supply Storage** Food and supply storage areas house most the products that dining services buys. The assessment team can try to identify overpackaged products or products purchased in very large quantities and talk over how various products are used. Team members who know how food is packaged and sold will be able to identify opportunities for bulk purchases, and the team as a whole can compare products and their packaging. Some unlikely products hold surprising opportunities. For example, switching from canned to dried bean products reduces waste cans. Changing from number 10 cans to a "bag-in-a-box" reduces the waste cans (the box is equivalent to the cardboard box that usually holds six cans, and the inner bag liner is much less material than the can).

The assessment team should check the walk-in refrigerators and freezers, discussing how these are used and examining their condition. The refrigerator door seals can be checked quickly by shutting a piece of paper in the door and trying to slide it out. If it pulls out easily, the door is wasting valuable energy. Clear-plastic-slatted energy curtains should cover the door to every refrigerator and freezer. These curtains should be used regularly and be in good repair. The assessment team can check the lights in the refrigerators to see if they are on all the time, wasting electricity and heating the spaces the university pays to keep cool. If refrigerator lights are on all the time, it may be because they require that the staff switch them on manually each time they enter the door. Adding switches that turn on when the door is opened can save money. Compact fluorescent lamps can also be installed in the refrigerators. Refrigeration and freezer space should be efficient, with properly sized refrigeration. During vacations dor other periods of low use, the contents of some refrigerators can be combined so that one or more units can be shut down.

**Kitchen and Food Preparation Areas** The kitchen and food preparation areas are the places for an audit team to observe how waste is generated and collected. Preconsumer food scraps can often be collected in the salad preparation area. The team should talk with the bakers and cooks about the way they use, warm up, and turn off their equipment. The team can observe ovens and steamers (Are they on although they are empty?). They can ask staff for ideas about ways to cut waste and reduce costs. Observing the general working conditions (Are they hot? crowded? new or old?) will help to identify opportunities where environmental actions can solve other problems as well.

**Serving Line** The serving line usually has steam tables (tables where hot water beneath the food is used to keep food warm), heating and cooling carts, or small refrigerators that may be able to be turned off on between meals. The types of plates, cups, flatware, and trays can be observed as well. If the kitchen uses disposable dishes, the team can discuss the potential for increasing the use of at least one durable product such as trays. Sandwich cooks and the staff who serve food can tell the assessment team their ideas about opportunities for waste reduction and efficiency. They can also observe if overhead lights, warming lamps, or lamps in display cases stay on all the time and if switches are accessible and in known locations.

**Dish Washing** The dish-washing staff can identify which meals generate the largest amounts of food waste. The team can observe how waste paper, bottles, food, and other materials are captured for disposal. What kind of storage space and containers are already available? They can examine the dish-washing equipment to see whether it reuses rinse water and whether faucets are left running or are in poor repair.

**Dining Room** In the dining room the team can think about how the customers receive and consume their food. Can dining replace small packets of sugar and salt with larger dispensers? Can dining substitute a bowl of yogurt for individual containers of yogurt? Does the location of the soda fountain prompt the use of multiple cups? How can the dining experience educate customers about upcoming environmental initiatives and their role in making them succeed?

### The Committee's Role in Dining Services Decisions for the Environment

Because university dining services is generally quite contained, a university committee will often not address dining issues except as part of broader university policy. A dining environmental committee may be a more effective mechanism for addressing energy, waste, and hazards in dining services. However, it is very important for a campus-wide committee to include representation from dining services so that their efforts are integrated with campus-wide efforts.

**Box 5.1**
Tufts dining services strategic plan introduction

**Environmental Stewardship Initiatives**

Dining Services needs to expand existing initiatives by developing procedures to reduce or eliminate adverse environmental impacts. Dining Services is committed to incorporating the guidelines of the Tufts Environmental Policy in its operations by educating customers and staff in reevaluating operating procedures to offset the adverse environmental impacts of our industry. We seek to play a leadership role in our industry and on campus. We will support other campus environmental initiatives.

### Establish a Dining Environmental Committee

Once the team has completed the initial walk-through assessment, it is time to form a working committee to address the specific environmental initiatives outlined in the remainder of this chapter. Dining services may want to have a committee that addresses issues across all dining units, as well as a working group for each unit. The committee should include dining managers, kitchen staff if possible, and customers. If the university has an environmental manager, energy manager, or recycling coordinator, they will be valuable contributors.

Tufts University used its dining committee to establish an environmental action area for inclusion in the department's strategic plan (see box 5.1 and appendix B). It then spent time addressing the details of individual actions. In another project, Tufts used a committee of cooks and other staff to achieve significant energy saving through turning off ovens, steam tables, refrigerators, and lights.

A dining environmental committee can be instrumental in making changes happen by providing valuable energy and enthusiasm, research, product and choices, links with other universities, and staff training.

### Working with a Contracted Dining Service

Many universities contract with outside firms to provide on-campus food service. Some of these firms are local, but many have national affiliations, such as Kraft and Marriott. Because they are under contract to provide a service or are managed from off-site, it is sometimes more difficult to work with them than with university staff performing the same job. Employees of a contracted dining company may feel that they are not at liberty to make changes in the suppliers they choose for foods, cleaning products, or other supplies without receiving approval from their managing company. As with a university-operated dining service, contracted dining services have many opportunities to reduce environmental impacts. In both cases, the success of the initiative relies more on the willingness of the managers and the staff than on the contractual arrangement.

The development and writing of a contract for a contract dining service company can be an appropriate place to interject specific and far-reaching environmental requirements without increasing demands on university staff or resources. The responsibility for negotiating and overseeing the contacts varies from university to university but usually falls to operations department administrators. Dining services contracts might specify the use of organic produce, recycling, and energy-efficiency standards. Despite the service contracted for, universities should link the costs of waste disposal, water, wastewater disposal, and energy to the contractor's operation and performance by billing it directly for these services. Doing so will link their consumption and daily use of resources with their real costs and can provide incentives for efficiency and waste reduction. Monitoring contracted service companies to ensure that their performance meets the contract and project goals is important for making the relationship a success for the environment.

## Reducing the Generation of Solid Waste

Although few comparative data are available, dining services in a residential undergraduate college can contribute 10 percent or more to a college's solid waste stream. Dining services generates solid waste as food waste, cardboard boxes, plastic containers and packaging, cans, glass jars, paper, plastic, and aluminum wrap. In addition, food packaging is energy intensive. A study showed that packaging is responsible for almost 20 percent of the energy use in the Swedish food production process.[1] Activities on the loading dock, in the kitchen, in the dining room, and in the dish room all generate solid waste. An estimated one-third of the solid waste in dining is corrugated cardboard, an easily recyclable packaging material. In the dining hall as across the entire university, reducing solid waste and reusing or recycling the remainder is the most effective way of dealing with it. Abatements in solid waste benefit the environment by reducing the use of natural resources, reducing the transportation of goods and their packaging, and reducing the material disposed of in landfills or incinerators. Dining services has many opportunities to reduce waste from packaging, food waste, and service waste. In general, these reduction strategies are cost-effective and can be easily institutionalized. Best of all, these strategies save money in procurement and in waste disposal charges. Table 5.1 shows some examples of waste reduction steps.

## Purchase in Bulk

One way for dining services to reduce the amount of packaging waste is to stop buying as much packaging. There are many products on the market that are available with less packaging and good quality. Most food comes in cans, bottles, jars, boxes, or other containers with secondary packaging such as shrink-wrap, pallets, or shipping boxes. Some of this packaging simplifies handling and ensures safety, hygiene, and freshness, but some can be eliminated.

Condiments such as cream, sugar, and ketchup are available in large bulk dispensing units. Harvard University dining services has made an extensive effort to change from portion-control packs of condiments to centrally distributed products. Purchasing condiments in bulk and serving them in reusable containers can save money since the per-serving cost is lower and there tends to be less waste. Implementing bulk containers of products used by the customers needs to be done carefully because loss (and waste) can be great if customers easily borrow, steal, or break these larger containers. Many university dining services departments purchase breakfast cereals in bulk and serve them in dispensing units rather than purchasing individual boxes; the larger containers reduce costs because it is nearly impossible for students to remove cereal from the dining halls for their own use in their rooms. Concentrated juice, stocks, and detergents that are diluted on-site are another effective way to buy in bulk.

Determining the products that are available in bulk is often difficult. Dining services can ask suppliers to help identify products that generate less. A letter to a supplier (see box 5.2) can explain that the institution

**Table 5.1**
Examples of waste reduction in dining services

| Product | Waste reduction action | Environmental benefits | Trade-offs |
|---|---|---|---|
| Kidney beans, garbanzo beans | Purchase dried in plastic bags rather than in steel cans | 90 percent reduction in waste | May increase labor and water use; steel can be recycled more easily than bags |
| Fresh produce (e.g., broccoli) | Purchase precut or prepeeled | Reduces food waste generated on-site | Assumes that food producers deal with food scraps responsibly (i.e., by composting or as animal feedstock) |
| Reusable secondary package (e.g., crate in which cans are transported) | Crate is reusable for same use | Generation of cardboard boxes is reduced | Storage and back hauling can be problems |
| Cereal or rice and other items available in large containers | Buy in bulk | Reduces the generation of primary packaging | Variety and storage can be problems |
| Products in small quantities for retail consumers—often offered at special prices in special circumstances for products such as meats | Consumer packs are individually wrapped and generate more waste than bulk pack | Avoiding consumer packages reduces waste generation and saves on employee time | May mean missing a seemingly "good deal" on these products; full cost of time and waste disposal change the equation |

**Box 5.2**
Sample letter to vendors

Dear Vendor:

Green University is committed to the protection of the environment in the education of our students and our business operations. Green University dining services has considered a number of ways that it can take action to meet the guidelines laid out in the Green University Environmental Policy (attached).

We have observed that your product [*product name*] is only available to us in two-pound packages. Since the reduction of solid waste is among our goals, we would appreciate it if you would consider making this product available in bulk 20- or 40-pound packages. This would reduce both solid waste and labor in handling this product.

In addition, please provide us with a list of products that are available in bulk. We are particularly interested in those that we are not currently using. Please keep us posted about environmental initiatives that your company or those you represent are pursuing.

Sincerely,
Director of Dining Services

is trying to reduce packaging waste and is taking other environmental initiatives. Dining services can use the letter to request a list of all items that are available in bulk and the price advantages of bulk purchasing.

When products are purchased in bulk, additional equipment may be necessary. For example, handling large, heavy bulk containers, such as vegetable oil in 55-gallon drums or 100-pound bags of flour or rice, can increase the risk of back injuries to dining workers. Appropriate lifting equipment for these heavy containers may be necessary. Suppliers can tell dining services if the bulk material requires special handling (such as a pallet-jack to handle extra weight) or dispensing equipment (such as dispensers for cereal or mixing units for juice) or storage.

### Use Reusables When Possible

The university's dining services must choose between utensils, cups, and plates that are reusable, such as metal forks or ceramic plates, or single use, such as plastic forks or paper plates. The choice often generates a good deal of discussion and debate. The development of viable programs to capture and recycle polystyrene cups, plates, and utensils has made this debate even more confusing.

Unless a university is located in an area with severe water shortages, reusable plates, cups, and utensils are the most environmentally sound choice for the university food service operations that have eat-in service. Outside the dining halls, dining services should also use reusables at catered events such as conferences or meetings held on campus. Some argue that the hot water, soaps, and energy associated with the production of and cleaning of reusables have a more severe impact on the environment than disposables. Others argue that disposables provide greater sanitation. These studies often are funded by manufacturers of disposable products that have a clear stake in the results, such as those produced by the Foam Packaging Council, an organization of the Polystyrene Packaging Council. Academics and public relations firms continue to argue the details of the case for each material. Reusables are usually equal to or better for the environment than disposables.[2]

Some colleges and universities have inadequate dish-washing facilities, designed when disposables were the accepted serviceware option. It is difficult to make a complete conversion to reusables without proper dish-washing facilities, which can be a major construction or renovation project. But dining services may be able to change some disposables to reusables or change to reusables during off-peak hours. For example, reusable trays could replace disposable trays. Similarly, dining services may consider using durable flatware, plates, and glasses during slower off-peak hours (close to opening or closing times when fewer students use the facility).

Replacing reusable china that has been "borrowed" for use in student apartments or academic departments or accidentally removed from the dining halls costs some universities thousands of dollars each year. At Tufts, eighty sets of flatware disappeared from one take-out cafeteria in three days! A few university dining halls deter this theft with deposit systems and inventory control systems. Other programs use student education campaigns and highlight the environmental benefit of reusables and the fact that theft makes disposables the cost-effective alternative. At Tufts, Patti Lee, director of dining, considered an amnesty program that would allow students to return china—no questions asked—in return for similar quantities of old, mismatched patterns that were of little use to the dining department. Unfortunately the program was never implemented.

Reusable dishware can be promoted by giving a travel mug to first-year students and new staff members. UCLA's refillable mug program kept more than 20,000 disposable cups out of the landfills each month while it was implemented.[3] Dining can promote the reusable cup program by offering discounts on beverages to those who bring their own mug. In one university program, the cafeteria charges for paper cups, and the posted price reflects the price of a beverage without a cup. Mug programs are paid for out of dining and recycling receipts, and dining departments use the mugs to promote their cafeterias or services. At other schools, student groups sell the mugs as fund raisers.

Concerns about sanitation of mugs and coffee pots or soda fountains potentially contaminated by contact with a dirty mug have discouraged some mug programs. One option is to offer a "free wash with a fill-up" to promote mug hygiene. Another possibility is to charge a deposit on a mug that is returned for another (clean) mug when students get food. Soda and coffee machines that do not require contact with the cup lip when filling can alleviate some concerns about the mug programs.

Reusable rather than other single-use items such as aprons, hats, and towels can save money and reduce waste. As a result of their environmental efforts, food service employees at the Lenox Hotel in Boston started wearing their own hats rather than paper hats. The change saved the cost of the hats, reduced waste, and gave employees an opportunity to have a voice in their uniform. This same change was problematic at one college, when waste-conscious kitchen employees started wearing hats from the college's main football rival!

### Reduce the Weight and Volume of Disposable Dishes

In almost every food service operation, some applications, events, or food service units require disposables for take-out uses. Pursuing waste reducing in the selection of disposables can dramatically reduce costs and waste generation.

Because of a partnership with the Environmental Defense Fund, McDonald's Corporation switched from polystyrene clamshells to lined paper wraps for sandwich products. The results included reductions in

**Table 5.2**
Comparison of alternatives for serving take-out

| Sandwich wrap option | Typical weight per serving | Typical cost per serving | Notes |
|---|---|---|---|
| Aluminum foil | 20 grams | Varies depending on weight | Very energy intensive to produce; poor choice |
| Heavy molded or stamped paper plate | 17 grams | $0.07 | Available from preconsumer recycled paper |
| Picnic-weight paper plate | 12 grams | $0.03 | Flimsy and unsatisfactory for many applications |
| Polystyrene plate | 12 grams | $0.04 | Recyclable in some programs; collection and storage are problems |
| Waxed paper plate | 10–12 grams | $0.03–0.05 | Lighter and less bulky than stamped plate |
| Waxed paper wrap with identifying sticker | 7 grams | $0.02 | Offers major volume reduction as well as cost savings |

energy consumption, air emissions, and water pollution during manufacture and four times less solid waste.[4] Table 5.2 compares alternatives for a take-out sandwich. The alternatives are ranked by weight, which also usually corresponds to their environmental benefit. When information is incomplete, the lightest-weight choice is most often the best alternative for the environment. In this case it is also cost-effective.[5] Aluminum foil should be avoided as a food wrap whenever possible, since its production is extremely energy intensive.[6]

Despite the financial benefits of changing from a heavyweight paper plate to waxed paper wrap for cold sandwiches, Tufts dining services was slow to carry out this change. The first barrier was the lack of counter space for the wraps, and so the buildings and grounds department made special vertical wrap holders. The second barrier was labeling the sandwiches so the cashier could identify them. To solve this, they purchased preprinted labels that also helped to tape the wrap closed around the sandwich. The more difficult barrier was the dining staff's perception

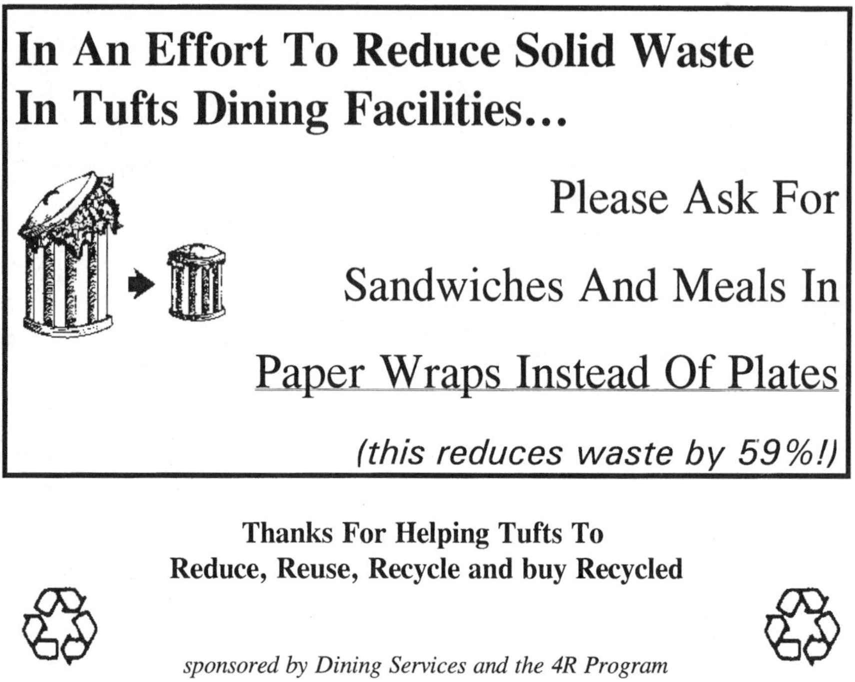

**In An Effort To Reduce Solid Waste In Tufts Dining Facilities...**

Please Ask For

Sandwiches And Meals In

Paper Wraps Instead Of Plates

*(this reduces waste by 59%!)*

**Thanks For Helping Tufts To Reduce, Reuse, Recycle and buy Recycled**

*sponsored by Dining Services and the 4R Program*

Figure 5.1
Sign used to encourage the use of sandwich wraps

that their delicatessen service was a first-class kitchen and that paper wraps denoted fast foods. To overcome this, the program appealed directly to the customers with the use of the signs, shown in figure 5.1. Customers began asking for the sandwich wrap, and the program finally succeeded in reducing waste and saving money. Nonetheless, the managers' preference prevailed, and after trying the wraps for several months, the managers finally settled on a lighter-weight plate and now use the wrap only for take-out sandwiches.

Other changes to reduce waste can be invisible to the customer yet have dramatic impacts. For example, McDonald's reduced the wall thickness and length of straws to cut down the amount of polypropylene in each. Because the company uses so many straws, the aggregate change is significant, although no data are available to measure it. Dining services can look for opportunities for waste reduction in their own operations.

**Evaluate the Access to Disposables**

Many dining facilities offer students and staff a choice of reusable or disposable cups. Eliminating the disposable option or making it more difficult can reduce waste. In contrast, many dining departments, including that at Harvard University, have found that they can drastically reduce the use of disposable napkins by increasing the access to these same napkins. Harvard Dining found that patrons took only the napkins they needed when napkin dispensers were placed on each table rather than in a central location.

**Recycle Remaining Solid Waste**

In college and university dining services, recycling is likely to be an important part of the environmental stewardship efforts even if dining services has pursued waste reduction aggressively. Dining services can recycle many of the materials that make up the waste stream in a dining service operation: cardboard, steel, glass, and some plastics. Furthermore, materials that are not recyclable can be replaced with recyclable alternatives.

Developing a successful recycling program is labor intensive, so dining services' recycling programs should be coordinated with other university recycling programs. In some cases, dining services may be able to take advantage of recycling programs that vendors offer. Storing material and negotiating to have vendors haul it back to their warehouses for recycling in their own programs is an alternative. Because dining services often has large quantities of materials collected in several locations, they can sometimes sell the recyclables at higher prices by consolidating the material at one location. Combining materials with those from other institutions can increase the marketability for a recyclable material. For example, before Tufts developed a metals recycling program, the buildings and grounds department hauled dining services' steel cans to the local community recycling center, which had a well established program. In return, the university brought the community's cardboard back to campus for inclusion in the university cardboard recycling program.

When looking to recycle wastes, dining services can begin by determining the types of waste generated and how these wastes are generated and handled. Although quantitative measures are appealing, qualitative measures, such as those gathered from looking in the storeroom and trash dumpster and talking with the dining staff, are often as effective and more efficient, at least initially. If more detailed data are needed, students or dining staff (or a team of the two) can collect and measure the solid waste material for a week. Purchasing records may help figure out the monthly average use of products in glass, plastic, or cardboard.

Unless the waste reduction efforts are superb, corrugated cardboard is likely dining services' biggest single waste item. Cardboard may account for 30 to 50 percent of the total dining hall solid waste. (Some of that waste is paperboard—material that resembles a cereal box. This is not corrugated cardboard and is generally not of value in recycling markets.)

Whatever the materials to recycle, dining services managers must consider and plan for the following:

- Collection and storage of recyclable material in the kitchen during the meal preparation
- Collection and storage of recyclable material between recycling collections (may involve volume reduction through crushing or compacting)
- Rinsing of recyclable containers to prevent odor and lessen the chance of attracting rats and insects
- Contamination accepted by the recycler (e.g., are labels allowed on cans and are caps allowed on bottles?)
- Availability of the market for the recyclable material (i.e., does anyone want the material?)
- Costs
- Waste reduction alternatives that reduce the recyclable material, such as purchasing items in bags rather than boxes

To develop a recycling program, dining services must work closely with the buildings and grounds department and others on campus who coordinate recycling. In addition, recycling companies, anxious to sell their services, can provide information about methods of handling material in the kitchens and on the loading docks. Local and state environmental officials and town or city recycling coordinators are resources for information about local requirements, new technologies, and links with others to increase the marketability of a product or improve the collection and storage techniques, find markets, and share costs.

New and innovative recycling services and products are entering the market every day, so dining services should not try to invent solutions to recycling problems such as storage, compaction, or handling. Students are excited about recycling and ready resources for investigating markets and recycling systems. Student projects can help to evaluate and compare a variety of systems to ensure that dining services is well served.

Table 5.3 lists the most commonly recycled materials in dining services. It also shows some of the issues specific to dining services that should be considered in recycling that material safely and successfully.

**Select Recyclable Products**

In order to make recycling easier and more cost-effective, dining services may want to find ways to reduce the need for recycling programs for certain materials by waste reduction and substitution. Some recyclables can be eliminated altogether by finding alternative packaging. For example, in order to reduce the safety problems associated with handling glass and the hassle of recycling small volumes of glass generated on campus, one college declared the campus glass free and eliminated the need for a glass recycling program. Since glass is difficult to handle, is usually of low value as a recyclable material, and is easily contaminated, other campuses may wish to consider a similar policy.

Some products are available in several different packages. At Tufts, recycling steel cans was problematic because of a lack of storage. Dining managers observed that the majority of the steel cans held tomato products, which are available in an alternative "bag in a box" package that was almost entirely recyclable (the plastic liner was discarded as trash) in Tufts' cardboard recycling program. The switch reduced the generation of steel on campus and helped solve several logistical problems.

**Buy Recycled**

Collecting and shipping cardboard, steel, plastic, and glass for recycling is not worthwhile if the recycled materials are not marketable. Purchasing products made from recycled materials helps to close the recycling loop. Dining services can ask vendors for products or product packaging made from recycled materials. The most effective way to do this is to specify recycled packaging in purchasing contracts. Products such as recycled corrugated cardboard and unbleached 100 percent, postconsumer recy-

**Table 5.3**
Materials commonly recycled by university dining services

| | Corrugated cardboard | Glass | Metals: steel, aluminum | Plastics |
|---|---|---|---|---|
| Most common sources in dining | Boxes | Beverages; other bottles | Steel: cans<br>Aluminum: beverage cans, foil, pie plates | Pickles, condiments; disposable dishes; other packaging |
| Other major sources on campus | Bookstore, warehouse, purchasing, library | Offices, dorms | Buildings and grounds; used equipment | Small quantities throughout |
| Recyclability considerations | Waxed cardboard, often used for vegetables, is rarely recyclable; paperboard not recyclable | Separation by color adds value; where a bottle deposit exists, bottles must remain intact | Cleaning and crushing cans is time-consuming and may be hazardous | Difficult to find profitable and reliable markets and difficult to clean |
| Methods of collection and storage | Loose and flattened; baled; crushed | Loose; crushed; cases | Loose; flattened; crushed | Loose; baled |
| Special considerations for dining services | Must be clean; dining generates large quantities | Broken bottles can be hazardous | Safety concerns in flattening and washing | Difficult to store; poor market |
| Waste reduction opportunities | Buy in bulk; return for reuse | Buy beverages in concentrate and dispense through vending machines | Buy in bulk or concentrate | Switch to more easily recyclable materials and reusables |

cled paper for take-out bags and napkins help to bolster recycling markets.

## Food Waste

In a dining service that uses reusable dishware, as much as another third of the waste is pre- or postconsumer food waste. The preconsumer food waste includes the outer lettuce leaves, unusable leftover food, and grease from frying food. The postconsumer food waste is food that is served but not consumed. The remaining third is other packaging and containers.[7] Talk to any dining services staff on any campus, and they will tell you that students waste a great deal of food. Students may tell you that they do not like the food, but the reality is that many meal plans provide "all you can eat" meals to students, encouraging them to try foods and increasing the incentive to waste food.

The problem of food waste exists outside campus cafeterias too. Food waste accounts for 9 percent of the estimated 195 million tons of municipal solid waste generated daily in the United States, or almost a half-pound per person per day.[8] A joint study by McDonald's Corporation and the Environmental Defense Fund found that 34 percent of the solid waste hauled away from the average McDonald's restaurant was organic food waste, such as coffee grounds and eggshells—most of it generated behind the counter rather than by the consumer.[9] Figure 5.2 shows the makeup of food waste at Tufts.[10]

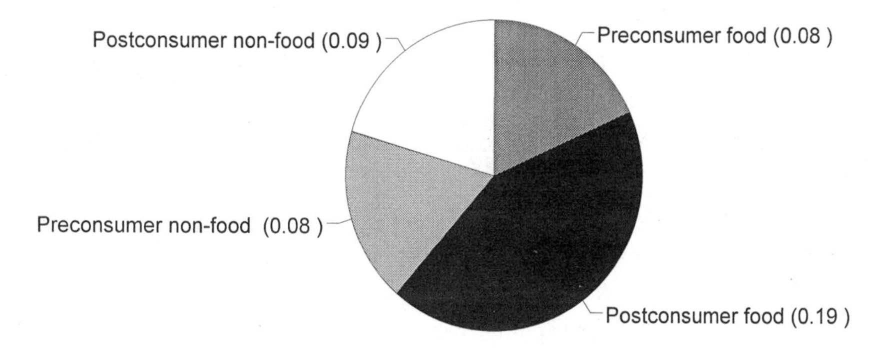


**Figure 5.2**
Waste per person per meal in Tufts dining halls (in pounds per person per meal)

### Reduce Food Waste

In university dining halls, as much as three to four ounces of food remain on the average plate after the meal (postconsumer waste). The food preparation steps probably generated another one or two ounces behind the counter (preconsumer waste). This may not seem like much, but a moderately sized university can throw out as much as a ton of food a day.[11]

Student culture and attitudes make food waste a difficult issue to curb. Using students to lead "take what you can eat" initiatives and focusing on the environmental benefits rather than just the financial costs can be effective. One university that set food waste reduction goals collected the food waste from student plates, displayed it outside the dish room, and measured progress with daily tallies.

A la carte food pavilions can also reduce food waste by providing a great deal of variety and requiring patrons to pay for each item they select. A program of trial size portions can also help to reduce food waste associated with food sampling. Dining services managers can take steps to reduce the postconsumer food waste by ensuring that portions are correct and menu planning is accurate. Purchasing foods that have had some preprocessing, such as broccoli that is cut and carrots that are peeled, can reduce preconsumer food waste on-site. If the suppliers of these prepared vegetables are composting the waste from their food preparation facilities, there will be a net reduction in total solid waste. (If they are disposing of the food preparation wastes as solid waste, the university simply moves the generation of waste from its facility to an off-site facility.) However, the larger volumes of food waste generated by a food production company make it more likely that they will be handling the food waste as resources rather than as waste (such as animal feed or compost).

### Donate Excess Food to Food Pantries

Donating excess food to food pantries and homeless shelters can keep leftover food out of landfills. Some donations may be made on a regular basis. Some schools find that they have larger volumes just before vacations or summer break, or after special events such as conferences or parents' weekend. The chaplain's office or community relations office can be helpful in establishing a regular program.

There are a number of logistical considerations with the donation of foods to shelters. Proper sanitation and food preservation and the associated liability to the institution are likely to concern many dining managers and deter some food donations. Many dining managers are justifiably concerned about the potential liability for food eaten off-site. They are concerned about a shelter's failure to refrigerate or reheat the donated food or to sanitize the kitchen and eating area. In most states, Good Samaritan laws offer protection from liability when good practices are followed. Developing an ongoing relationship rather than a sporadic program can help the university to understand the shelter's food needs. An ongoing relationship helps the shelter to plan around the university's food availability and arrange appropriate transportation and food storage. If necessary, a university's legal counsel can draft a waiver of liability for food eaten off premises.

**Turn Food Waste into Animal Feed**

Historically, institutional food waste was collected and fed to swine or cattle. As suburban sprawl pushed farms farther from the urban centers and transportation costs increased, farmers began to charge for the hauling service. In addition, in most states, farmers are required to sterilize postplate food wastes fed to pigs in order to eliminate the risk of trichinosis. The result is that feeding grain is easier and sometimes less expensive than food waste. Nonetheless, the rising cost of solid waste disposal may make hauling food waste cost-effective again. State departments of food and agriculture can help locate interested farmers who raise pigs and cows.

The school calendar can be problematic for farmers if the college is the sole source of food waste. Combining efforts with a local hospital or nursing home that will have a steady source of foods can make a farmer's trips more worthwhile and reduce the cost of pickup.

Efficient and sanitary collection and storage of food waste often require some adaptation of existing waste handling methods. Appropriate containers, usually wheeled and with tight lids, both inside the kitchen and on the loading dock or storage area, are important for success. Arranging for frequent food waste pickups (daily if possible) and rinsing the containers between loads can help prevent rodents and flies. If cool or refrigerated storage is available, dining services can store food waste for longer periods between pickups (up to a week). However, health codes usually dictate that waste be stored separately from food supplies.

**Compost Food Waste**

Composting is a biological process that breaks down organic matter, including food waste, into valuable material for improving the soil quality in gardens and fields. Most food waste is compostable and can provide needed nitrogen to carbon-rich leaf waste found in most composting operations. Dartmouth College and Hampshire College both compost food on-site with their yard waste and animal manures. Ithaca College also has an extensive food waste composting facility with a dedicated building.

Composting food waste can be trickier than composting leaf and yard waste. Preconsumer vegetable waste, such as carrot peels and broccoli stems, can be combined with leaf and yard waste without a problem; universities that manage their own composting facilities may, in most states, add preconsumer food waste to it. Postconsumer wastes, however, may contain meat scraps, bones, and paper (napkins) that do not compost as quickly or easily and may attract animals to the composting facility. Cooked meat products are the most difficult to compost since they often have little surface area for bacteria to reach, they can attract animals, and they are smelly as they decompose. Food handling systems that grind or pulp food, common in many dish rooms, can grind the food waste, providing sufficient surface area to make the material easily compostable. Composting postconsumer food wastes is also often regulated differently from composting leaves or preconsumer food, and composting them in some states is considered experimental.

If the university does not have a yard waste composting operation on-site, dining services may still want to try composting. Small quantities of food waste (several gallons per day) can be composted in backyard tumbler systems. These backyard systems have worked well at small elementary schools where students, under supervision, monitor and turn the compost and use the final product in on-site gardens. Backyard systems can work successfully at small student residences (those with twenty-five or fewer rooms) with their own kitchens. In either case, staff or students will need to be on hand to manage the composting operation. University kitchens are likely to generate larger quantities of waste than backyard systems can reasonably accommodate yet smaller volumes than justify large-scale commercial composting equipment. Most colleges and universities will find hauling their food waste to a commercial composting operation or arranging for a pickup is most feasible. These commercial operations handle waste from many generators and then sell the finished product for a soil amendment. For example, Bates College in Lewiston, Maine (resident student population about 1,400), hauls over 40,000 pounds a year of preconsumer waste to a local farmer who composts the material and then returns the finished compost to the college grounds.

The composting industry is a burgeoning one. It is likely that entrepreneurial commercial companies will begin to see the value of institutional food waste to improve the quality of their compost and may provide collection and transportation services of waste material to their sites. The solid waste division of the local department of environmental protection can provide information about nearby commercial composters.

**Keep Food Waste out of the Drain**

In many university dining services kitchens, industrial-sized garbage dispose-alls grind food waste and send it off to the local wastewater treatment plant. Many large wastewater treatment systems now prohibit this practice, or will soon prohibit it, since the increased organic matter is problematic for the sewage treatment process. Universities that handle their own wastewater on campus find that a surprising amount of their treatment capacity is needed to accommodate the treatment of food waste. Feeding the waste to animals or composting it are two alternatives to washing the material down the drain. When composting or a livestock farmer is not available, disposal of food waste as solid waste is more prudent and often more cost-effective (because of the high cost of water) than flushing food down the drain.

## Kitchen Equipment

The energy used to run kitchen equipment has both environmental and financial costs. Although energy costs in dining (estimated to be about thirteen cents per meal[12]) are only a fraction of dining's labor and food costs, the dining services energy can amount to 10 percent of total university energy costs. Since dining is such a large and concentrated user of energy, usually natural gas and electricity, there are significant opportunities for savings.[13] Selecting and maintaining kitchen equipment to ensure maximum energy efficiency and carrying out low-tech efficiency measures can reduce energy use.

Training and attention can help dining services staff serve quality meals on time that are prepared and served with the minimum amount of energy. Even dining halls with state-of-the-art efficient stoves, ovens, and other equipment can benefit from incorporating some low-tech strategies. Facilities with older equipment can realize savings of several thousand dollars annually.

Successful energy-efficiency efforts begin with a commitment from the dining director in cooperation with the buildings and grounds department. The director and his or her managers may be unaware of the impact that the energy used in the dining halls and kitchens has on the environment and the potential to reduce these impacts, as well as the financial costs. Because dining staff and customers are far removed from the energy costs, they may even operate as if energy were free. Thus, an energy-saving campaign in dining services should include education of all members of the dining services department. Dining services, or the dining environmental team, can design programs for each campus dining hall individually. In order to find successful strategies for reducing energy use, the program should be tried in a pilot location.

### Conduct a Walk-Through Assessment

An in-depth examination or assessment of dining facilities in order to examine equipment and how it is used will add detail to the initial walk-through described earlier in this chapter. This assessment can be used to identify all energy-using equipment, evaluate its condition, and determine how it is used. The assessment step should include participation from the kitchen staff and dining unit manager, as well as a representative from the buildings and grounds department.

Asking staff how equipment is used—when it is needed and when it is not—is the most important data-gathering step. Their explanations should be confirmed with observations. For example, staff may say they turn steam tables off between meals, yet observing the serving line may indicate that this only happens half the time. Conducting several walk-throughs at different times of the day and week can help catch problems and identify trends.

An assessment of kitchen and dining hall energy use does not need to be highly quantitative, although detailed information makes program evaluation easier. Because the predominant message in an energy program is "Turn it off," the length of time that a piece of equipment is on is as important as the amount of its energy demand. Use an assessment worksheet, such as the one in appendix C, to record the important information. If the dining department has a complete equipment inventory, the audit can begin with that list. Information about the equipment's energy demand and regular use patterns can be added to the database, providing a complete equipment profile.

If they are available, historical records of energy use in dining halls are useful for identifying trends. For example, you would expect to see lower electricity and gas use during vacation months when the dining halls are not in use. Accurate records can show which halls and kitchens are more efficient (per meal). By comparing the energy used per meal in various dining halls, dining managers can identify the most inefficient dining halls. Regular monitoring of the utility bills can help to identify problems such as water leaks or other sources of waste. Unfortunately, aggregate water, electric, and natural gas data are not always available in the best form. Often universities do not measure dining facilities' energy and water use separately from the rest of the building or nearby buildings. Where possible, install separate meters to measure progress and ensure that billing is accurate. When this is not possible, managers should carefully estimate and document the savings that result from the efforts. (It is often difficult for the dining department to receive financial compensation for these savings.)

### Select Efficient and Appropriately Sized Equipment

The efficiency of commercial kitchens varies widely, depending on the equipment's age, fuel, design, and size. New technologies are more efficient, faster, and easier to use than the standard stoves, ovens, steamers, and fryers found in most institutional kitchens. Not everyone will be able to redesign an entire kitchen, but it is important to consider energy costs when upgrading old equipment or buying new.

When dining services staff shop for new equipment, they should select the correct size for the use by assessing what the department plans to cook (rather than a measure compared with existing equipment). In order to conserve energy, dining directors and managers should avoid the temptation to buy more capacity than is needed since the equipment will cost more to run and maintain. Some new equipment may be multipurpose or provide increased efficiency. For example, new infrared technology can cook foods more quickly than a microwave oven using just 20 percent of the energy a conventional oven requires.[14]

Purchasing staff should examine the energy efficiency of the equipment considered, evaluating the energy required to do the job rather than over time. Natural gas systems used to be considered more efficient and easier to use than electric appliances, but research is demonstrating that a number of new electric technologies can perform more efficiently than gas.[15] Induction cooktops, for example, provide immediate heat to the food and remain cool to the touch during cooking. Induction deep fryers also heat quickly and with higher quality results.[16] Some convection ovens reduce cooking time while using less electricity.[17]

Ventilation systems are costly to install and to run. Dining services should try to arrange kitchens to reduce the ventilation needs and consider the ventilation demands of new equipment. By grouping equipment with similar demands together, a kitchen can downsize some of the ventilation hoods. New ventilation technologies that use variable-speed controls rather than continuous venting save energy and may be appropriate in kitchens.

### Turn Off Lights and Equipment When Not in Use

Low-tech solutions such as turning off equipment when it is not in use and periodically checking and repairing equipment are highly effective. Because institutional kitchens use large ovens, dishwashers, and refrigerators frequently, significant financial savings and pollution prevention can result from simple changes. Low-tech solutions can be incorporated into existing operations. The key to success in implementing these options is to make employees aware of them and to provide rewards and incentives for carrying them out. Contests and competitions can be useful for starting a program, although they must be combined with ongoing training and reminders in order to sustain the effort.

In many kitchens, the employees turn equipment on when they arrive in the morning and turn it off after serving the last meal. They learn this practice from coworkers or develop it to ensure that steam tables and ovens are warm when needed. The most common reason given for leaving on kitchen equipment such as ovens and steam tables is that the equipment takes a long time to warm up. In actuality, the problem may be that the warm-up times of the oven or table are not well known and not considered in planning the meal. Dining services managers can measure the warm-up time and post it near the equipment so that kitchen staff will turn equipment off when it is not in use and turn it on in time to come up to temperature. If the kitchen staff uses detailed food preparation instructions, the instructions can include steps indicating when in the process to turn equipment on and off.

Although energy use varies by oven, stove, steam table, or refrigerator, dining managers and staff must learn a consistent and simple message: turn it off when not in use. Managers may need to make modifications to the equipment or its switch to simplify this effort. For example, staff can turn a small freezer off during vacations by unplugging it; however, if the outlet is located under the unit, it will be extremely difficult to accomplish this routinely. Purchasing an extension cord or installing a switch can make it easier to switch off the unit. This change will usually pay for itself quickly. Walk-in refrigerator lights should not be left on all day. The energy lost lighting empty space is wasted, and the heat gain to spaces being cooled costs money too. Table 5.4 shows some examples of equipment that can easily be turned off between meals and during vacations.

Dining halls are frequently empty between meals, so there is little need for full lighting. A practice of turning off lights saves a few dollars each week and sends a clear and visible reminder of the energy-saving efforts to the kitchen staff. When a dining hall has ample daylight, the lights can be left off during meals as well. Occupancy sensors or daylight sensors, where appropriate, may help in this effort.

Equipment that is no longer used or is used infrequently exists in many university kitchens and dining halls. Paying to keep an unused water cooler running or burning gas for a pilot light in an old and rarely used

**Table 5.4**
Equipment used in dining services

| Equipment | Opportunity to turn off |
|---|---|
| Steam table (2 tray) | Between meals |
| Stack ovens | Before and after they are needed |
| Convection ovens | Before and after they are needed |
| Freezer: 4 cubic feet | Meals when this freezer is not in use or full; vacations |
| Coffee maker: large four-burner | Between meals; carafes can be used to hold coffee in slack periods |
| Small refrigerator: 4 cubic feet | Vacations or when not in use |
| Lights: fluorescent and incandescent lamps | When room is vacant; when sufficient daylight exists |
| Pilot lights on stoves and ovens | Vacations |
| Personal computer | At nights and periods of idleness greater than an hour |

oven is a waste of money and natural resources. By identifying this equipment and turning it off, unplugging it, or even removing it, dining services can save. When the equipment is likely to be replugged in, dining services can post signs explaining the energy program and showing that the equipment is unused. Dining kitchen staff should be made aware of the audit of unused, energy-using equipment.

Each dining hall and kitchen should implement a vacation shut-off plan, detailing how food will be consolidated during extended vacations so that large refrigerators can be shut off. The plan should include staff time for consolidation and cleaning of the refrigeration systems. Shutting off a midsized (25 cubic foot) commercial refrigerator for one week can save about $18 in electricity. Smaller units in satellite locations can usually be turned off easily during breaks. For extended breaks, it may be prudent to extinguish pilot lights on stoves and ovens.

### Improve the Efficiency of Lights

Dining services managers, directors, and the dining environmental team can work with the buildings and grounds department or energy managers to upgrade the lighting in kitchens, serving lines, and dining rooms with energy-efficient lighting (see chapter 3). In addition to installing efficient lamps and ballasts, there are opportunities in most halls to rewire the light switches so that banks of lights switch on and off separately rather than all at once, providing significant savings in dining rooms where all the lights are not needed because of sufficient natural light. Compact fluorescent lamps with electronic ballasts are effective in refrigerators and over stoves. (Compact fluorescent lamps with magnetic ballasts are less effective because they are slow to come on, especially at high and low temperatures.) Replacing incandescent lighting with compact fluorescent lights in a refrigerator can save lighting energy and reduce the heat from the lamp. Tufts dining services found that it had to experiment with different compact fluorescent lamp manufacturers in order to find one that fit into existing safety shields covering lights in food areas.

**Keep Refrigerator, Freezer, Oven, and Loading Dock Doors Closed**

Each time a refrigerator door is opened, cool air is lost, and oven doors and doors to outdoor spaces can cause dramatic heat loss or gain to the indoor kitchen space. To the extent possible, all of these doors should be kept shut. Energy curtains, usually of clear plastic strips hung in refrigerator and freezer doors, can reduce heat loss from this equipment. Energy curtains are also often useful in outside doorways, where winter cold or summer heat are problems.

Some employees are reluctant to close doors or use the energy curtains since these practices are cumbersome and awkward, especially with full hands. Dining services' training for new and returning employees and regular staff training should include instruction to keep doors closed except when passing through them. Because dining services is usually a hierarchical management structure, it is particularly important that dining managers follow the same practices expected of the kitchen and other staff. In addition, dining services may want to provide incentives and rewards for continued observance of these energy-saving practices.

**Keep Equipment Clean and in Good Repair**

Keeping kitchen equipment in good repair can lengthen its life and improve its operating efficiency. Dining staff or service contractors who maintain the kitchen equipment should regularly examine the equipment for leaks in door seals and leaking coolant, poor thermostat control, and inefficient operation. Refrigerators and freezers, in particular, are more efficient when their coils are cleaned frequently—as often as every three months—to remove dirt and dust.

Dining services or their contractors should maintain the seals on refrigerator and freezer doors to ensure that they keep tight connections. General wear and dirt can cause leakage that may cause the compressor to run unnecessarily. The seals on oven doors should also be checked frequently for leakage. Dining managers can regularly check seals by closing the door on a piece of paper; if the paper falls out or pulls out easily, the unit is not well sealed and should be repaired. Managers should also calibrate the temperature settings on refrigerators, ovens, and steamers regularly to ensure that food cooking and preservation are accomplished without undue waste. Lighting also benefits from regular cleaning since grease and dirt may decrease light output by up to 40 percent.

**Fix Leaking Refrigerants**

The Clean Air Act regulates CFCs and HCFCs found in most refrigerators and freezers and makes the university legally responsible for complying with its leak repair provisions. University staff technicians must also observe the no-vent and recycling regulations when servicing refrigerants in their own equipment. University staff and outside technicians must become certified and use certified equipment. A regional EPA office can supply information. (See also chapter 3.)

**Reduce Energy Use Through Behavior Change**

The Tufts Dining Energy Efficiency Program, a joint effort of Tufts dining services and the Tufts CLEAN! Program, was designed to reduce energy consumption at the dining halls through behavioral and small-scale, low-cost technical changes.[18] The program established an energy efficiency effort at the Medford campus Dewick/MacPhie dining hall between January and April 1994 with the following goals:

- Reduce energy (electricity and natural gas) consumption at Tufts dining services.
- Reduce energy costs at the university.
- Increase awareness of energy efficiency among dining hall employees and students.
- Influence energy consumption behaviors of dining hall employees and students.

- Determine whether an energy-efficiency program could work at all the dining halls.

The first major step in the planning of the energy efficiency program was the implementation of an energy audit, designed and carried out by Tufts CLEAN! staff, in cooperation with the managers and representative employees of each of the five dining halls. An audit form was designed to obtain information pertaining to the types and locations of all items that consume energy (specifically natural gas and electricity), the amount of energy demand, the extent to which each item was used (e.g., all day, two hours each day), and opportunities for savings based on observations and discussions with dining hall employees.

Actual audits were conducted at each of the university's five Medford campus dining halls. The audits identified several major areas where opportunities for increased energy efficiency existed: lighting, refrigerators and freezers, and cooking and warming equipment. Within each of those categories, specific opportunities were identified, such as:

- Replacing incandescent bulbs with compact fluorescent
- Turning lights off between meals and when sufficient natural light exists
- Using cooking and warming equipment more efficiently
- Turning off all cooking and warming equipment when not in use or between meals
- Replacing loose-fitting seals on all refrigerators
- Shutting off all small refrigerators during vacations

In the model program, a meeting with the majority of employees in the target dining hall was held several days before the beginning of the spring semester. Although employees were already somewhat familiar with Tufts CLEAN! staff and the energy program, this initial meeting was designed to introduce the model program concept, review environmental impacts of energy consumption and the importance of energy efficiency, and cooperatively develop a number of efficiency strategies that the employees would feel comfortable following.

In addition, the project formed an energy efficiency committee, made up of employees from each of the dining hall areas (kitchen, dish room, storage area, serving line, and general area). This committee met periodically—once a week to start—to review the program and continue developing energy efficiency strategies for the program. As a result of the initial meeting, a number of suggestions were agreed on by both Tufts CLEAN! staff and dining employees as the top priorities for initial efficiency strategies.

Tufts dining services staff are very conscious of their obligation to provide quality meals on time. Because any proposal that threatened this goal could not be successful, the most workable actions relied heavily on the dining staff.

Based on the energy audit and discussions with Dewick/MacPhie staff, a number of energy efficiency strategies were implemented at the dining hall during January, February, and March. These strategies are shown in box 5.3.

In addition, numerous signs were placed throughout the dining hall to remind employees to shut off equipment and lights and use energy wisely.

**Box 5.3**
Strategies to reduce energy in a Tufts kitchen

- Turn off dining area lights when not needed during the day because of natural light and/or between meals.
- Turn off lights in serving areas between meals.
- Turn off all other lights when not needed (e.g., bathrooms, dish room, storage area, salad area).
- Replace thirty-one incandescent lamps (75 watt and 90 watt) with compact fluorescent (20 watt) in walk-in refrigerators, above stoves and ovens, and in the MacPhie serving area and dish room.
- Replace two 90-watt incandescent bulbs above salad bar with 25-watt lamps.
- Purchase extension cords for small ice cream freezers on serving line in order to facilitate unplugging the freezers for 160 hours (out of 168 hours) each week when they are not in use.
- Keep refrigerator doors shut.
- Check seals on refrigerators and replace old or loose ones.
- Brush or vacuum refrigerator coils regularly.
- Consolidate foods from smaller refrigerators into the walk-ins and unplug smaller refrigerators during holidays longer than a few days.
- Unplug the old water fountain (not needed).
- Test cooking equipment warm-up time needed and post results.
- Evaluate the day's menu and turn on cooking equipment closer to the time needed rather than first thing in the morning.
- Turn off steam tables between meals.
- Turn off large coffee machine between meals.

These signs were made more secure by laminating them. In late March, a "whiteboard" was posted in the kitchen with a list of energy efficiency strategies reminding employees of the tasks they had agreed to follow. The signs and the whiteboard are important reminders to both the dining staff and the students.

As a result of the strategies implemented, we estimated annual savings in the first year of $800 to $1,300 with the potential to save nearly $3,000 (cumulative) in two years in the target dining hall alone. If these efforts were implemented in all of the Tufts dining halls on the Medford campus, the savings could exceed $10,000 over two years. Perhaps as important, dining staff indicated that they had saved money—$30 per month in one case—at home as well by taking energy-saving measures.

## Conserving Water

The largest use of water in dining services is for washing dishes. Water is also used for cooking, food preparation, and ice. It is wasted in obvious ways, such as running faucets, and in hidden ways, such as leaks and dishwashers. The costs of water purification and wastewater disposal are driving water costs up in most areas of the country (and worldwide), and supplies of potable freshwater are dwindling and increasingly threatened by pollution. University kitchens and dining halls present many water conservation (and money-saving) opportunities in dish washing, food preparation, equipment use, and maintenance.

Water conservation makes sense from a resource conservation and a financial perspective alone, and it has the added benefit of saving energy used to heat water. The Electric Power Research Institute estimates that sanitation, mostly from dish washing, accounts for 15 to 20 percent of the energy used by a full-service electric commercial kitchen.[19] Although most university kitchens do not heat hot water with electricity, energy use by hot water and sanitation is generally great.

### Conserve Water When Washing Dishes

The dish room is the place to conserve the most water in the dining halls. Many institutional dishwashers recycle the water from the rinse cycle into subsequent prerinse cycles in order to reduce water use, so dining managers must be sure to purchase these models when buying new equipment. Filling dishwashers to capacity for each load can reduce unnecessary use. As in any other kitchen, in institutional kitchens many items are washed by hand. To conserve water, pots and pans can be presoaked to loosen food and washed by filling a basin rather than running the water continuously. Again, dining staff training should include water-conserving methods during dish washing.

Garbage dispose-alls use large amounts of water. Dining services should reduce their reliance on dispose-alls by separating garbage for composting or animal feed or even by disposing of it as solid waste. Reducing reliance on garbage dispose-alls is an important way that dining can reduce the water used in the kitchens.

**Install Faucet Aerators and Stop Leaks**

Reducing the flow from faucets in the kitchens can save water. Dining services managers, employees, students, or the environmental team can check the rate at which water flows from each faucet in the kitchen by timing how long it takes to fill a one-gallon container. Faucet aerators can be installed to reduce the flow to about two gallons per minute. If the flow from the faucet is much more than this, a reduction can save gallons of water for jobs such as washing hands, rinsing produce, and washing dishes. University kitchens may want to be able to have the option of more rapid flow to fill large kettles or steamers, which can be accomplished by installing a hose that bypasses the aerator.

Because a dripping faucet can waste seventy-five to a thousand gallons of water weekly, dining services should ensure that each kitchen has an established program for staff to identify leaks or drips. Managers can then coordinate with the buildings and grounds department or the plumbing staff to ensure timely repair of leaks, often most effectively identified and reported by kitchen staff.

Dining services can also adjust ice machines, coffee machines, juice or drink machines, and water fountains to reduce water use. For example, ice machines can be adjusted to provide appropriately sized servings of ice rather than an overflowing cup. Locating ice machines away from heat sources (e.g., ovens) will also reduce losses due to melting and the energy needed to keep the ice from melting.

**Practice Water-Conserving Habits**

Water waste often occurs when faucets are left running while a staff member searches for a pot or reaches for the telephone. Dining staff should turn faucets on as needed and off when they are not needed. Simple steps to minimize water use during the preparation of foods include defrosting foods in pans of water rather than in running water. Cleaning foods with low flows of water or in basins of water can also reduce the total water used.

**Evaluate Water Quality**

Federal and state drinking water regulations are designed to ensure the safety of our drinking water supplies. In general, this country's water supplies are the best in the world and should be safe from contamination such as sodium and bacteria. However, recent problems with metropolitan water supplies suggest that it is prudent for institutions to have water tested regularly for bacteria, sodium, lead, and other metals. Most universities send water samples to off-site laboratories for these tests.[20] In some universities, water testing is the responsibility of the environmental health and safety officer or department.

Lead is a particular health hazard for young children and pregnant women, but no one should ingest it. Lead in drinking water is common in some regions of the country where the pH of the water leaches lead out of the lead solder used in the central and localized water distribution systems. Lead-free water bought from a municipality does not mean that university water is lead free; lead can enter drinking water from on-campus plumbing. Universities should pay particular attention to lead levels in all buildings, including university kitchens, by testing the water at an established laboratory.

Increasingly, university offices, dorms, and kitchens are using bottled and filtered water because of perceived dangers of drinking tapwater. To eliminate bacteria, municipal water suppliers most often treat water with chlorine, which can leave a taste. Furthermore, tapwater, especially in summer, is often tepid, and Americans have come to expect ice cold drinking water. These factors too are making filtered and bottled water increasingly popular. Dining services should evaluate both filtered and bottled water carefully. Both options can be costly, and money may be better spent to alleviate the source of the problem. Typically, water quality does not pose a health problem that requires using bottled or filtered water.

If water filters are used, they must be maintained according to the manufacturer's instructions to avoid bacteria accumulation or other problems. If bottled water is purchased, it is important to verify with the supplier that the suppliers comply with all applicable FDA regulations (required only of suppliers that sell water across state lines). Because bottled water is regulated as a food (by the FDA rather than by EPA), it may have higher levels of sodium than are allowed in municipal water supplies. The transportation and bottling of bottled water are resource intensive and should be minimized if possible (e.g., by selecting local sources, if at all).

## Selecting Foods for Low Impact

The most common agricultural practices in the United States involve the use of large amounts of fertilizers, pesticides, and intensive land use practices. The resulting soil erosion, pests that can resist chemicals, lack of biodiversity, and surface water and groundwater contamination are virtually unnoticed by both the eating public and those who serve them. Furthermore, the long-term effects of diets that result from modern agricultural practices, such as the use of pesticides and hormones, are essentially unknown.

College and university food services have an opportunity to use their menus to improve health, reduce environmental impacts, and educate students. For example, purchasing foods from local farmers and suppliers can reduce transportation and the associated emissions. Using organic flour or produce can reduce customer exposure to pesticide residues, as well as off-site pesticide exposure and use. Increasing access to and the variety of vegetarian menus reduces reliance on meat, often raised with intensive farming practices that increase soil loss. Reducing meat in student diets also helps reduce the intake of fat and chemicals that accumulate in fats. Careful selection of fish can avoid promoting practices of overfishing and the consumption of mercury.

Making changes in food staples and the recipes that make up food offerings is time-consuming. Incorporating organic foods and increasing the diversity and availability of vegetarian choices are possible throughout a university's campus food service operations and menu items. Many dining services departments find that it is most practical to begin on a limited basis. As with decisions that affect solid waste, energy efficiency, and water conservation, a decision to reduce the environmental impacts of food selection and menu development requires examining a multitude of individual products, ingredients, and recipes.

The nature of a university's own food services will determine how these changes can be incorporated effectively. For example, in universities where the food services are managed by an outside contractor, university administrators can use the contract process to specify an increase in local, organic, and vegetarian food choices offered in the dining halls. When dining halls offer a food station approach that gives students a la carte choice of burgers, pizza, sandwiches, entrees, and salads, there are opportunities to offer organic, local, and vegetarian choices. Students are ready resources to assist dining services managers in researching and testing targeted products. For example, Tufts used student projects and a student intern to research the availability of organic foods, pilot-test their acceptability, and publicize the changes.

A switch to increased use of organic and vegetarian foods has environmental benefits, but successful implementation may require that individuals change some of their eating habits and expectations. For example, an increase in vegetarianism is successful only if the customers eat the increased vegetarian offerings. University dining services can play a role in these individual decisions by providing patrons with information about their food choices and their impacts.

**Buy Local Products**

Buying products from local markets saves on transportation and provides visible support for farmers and other vendors in the university's community. In the late 1980s, Hendrix College in Arkansas undertook an extensive study to determine the sources of a number of its food products. As a result it increased the percentage of its in-state food purchases from 6 percent to 30 percent of its total food.[21] The Hendrix College Local Food Project combined the resources of students, state officials, and local farmers to support local farmers and improve food quality.

State departments of food and agriculture can help university dining services identify the major food crops that are grown in the state and when they are available; the list often has some surprises. Matching this list with the products that are bought regularly helps determine the local products that can be substituted directly into the menu. Many institutional kitchens buy from large food distributors but farmers often form cooperatives that may be able to sell directly to universities without going through a middleman. For example, if guaranteed a steady demand for potatoes, carrots, or apples, a single farmer or a farmers' cooperative may be able to offer products at competitive prices.

Certainly one of the barriers to buying local products is that seasonal availability does not always coincide with the demand throughout the school year. Schools in the Northeast will not be able to find a variety of fresh produce for salad bars during most of the school year. However, other vegetables, such as squash, potatoes, and carrots, are grown locally and can be stored for long periods of time. Bates College in Maine buys produce from local organic farmers and stone-ground grains from a local mill when they are available. Some items like herbs and potatoes are available nearly year-round.

**Increase the Use of Organic Foods**

In 1988 the Surgeon General's Report on Nutrition and Health estimated that as many as 10,000 cancer deaths annually could be caused by chemical food additives. Increasing purchases of organic and natural foods or foods raised using integrated pest management techniques and free-range and natural farming can reduce exposure to pesticides, herbicides, hormones, and antibiotics. There are many foods that can be purchased organically, hormone free, or "all natural." In most states these terms are regulated and standards are available, although some farmers using low-impact methods find the standards cumbersome and choose not to become registered. It may be possible to have students and a dining team evaluate these farms and their practices and support them.

Dining services should look for produce and grains that are grown without pesticides, herbicides, or chemical fertilizers. They can also find meat, chicken, dairy, and poultry products that are raised without subtherapeutic levels of antibiotics (some "all-natural" farmers use antibiotics to nurse sick animals rather than administer them on a regular basis), hormones, or other unnatural additives. To find sources of these foods, ask a local cooperative extension service or department of agriculture. Food cooperatives may be able to provide the college or university with bulk foods on a regular basis. Large food distributors will buy organic foods too if their customers clearly specify them.[22]

Some organic foods cost more than those that are not grown organically. Dining services may need to increase meal charges to reflect an increased use of organic products throughout the menus or to find ways (e.g., through waste reduction) to offset these costs. Another way is to offer these foods only in cash operations, where they can be individually priced to reflect the higher cost. Because a university purchases large quantities of food, dining services may be able to work with a farmer to guarantee a market or to take organic produce that is bruised or damaged and not available for retail sale. These foods can be used in soups or casseroles at competitive prices.

**Go Low on the Food Chain**

Because some meats are grown and raised using intensive production methods and are fed vast amounts of grain, meat production is especially resource intensive. The higher a food is on the food chain (beef and pork are higher than poultry, poultry higher than fish, and fish higher than vegetables), the more energy and water are used.[23] In addition to being more resource intensive, beef is high in fat, making it a less healthy form of protein than alternatives lower on the food chain. Although fish is low on the food chain, some of today's fishing practices are highly disruptive to lakes and oceans.

University dining services alone is not likely to convert many students or other customers to vegetarianism. Indeed, some would argue that dining services should not take a role in personal eating choices, but in fact, any available menu does make some choices by what it offers and excludes. Providing interesting and tasty vegetarian menu choices can go a long way toward increasing vegetable consumption and decreasing the consumption of meat-based meals that are water, soil, and energy intensive. Increasing vegetarian meals may reduce the consumption of pesticide residues, artificial hormones, and herbicides.[24]

Finding vegetarian menu choices that are suited to the demands of institutional food service and are not always cheese or pasta can be challenging. Tufts dining services held a recipe swap and invited students and other customers to bring their favorite vegetarian entrees and recipes for others to try and share. The result was a collection of new recipes and a chance to test-market them. Decreasing the size of a meat serving by reducing the portion size or by substituting beans for a portion of the ground beef in a casserole is another way of decreasing meat consumption. As with any other change, customers need to be informed about the change and the reasons for the change if it is to be successful.

## Conclusions

Students visit the dining halls more often than they do any single academic class. Therefore, university dining services departments have a role to play in the greening of the university and the environmental literacy of the university community. The large contribution of dining services to waste generation and university environmental impacts give both opportunities and responsibilities to the dining department. The selection and preparation of foods and handling of wastes all offer ways for dining services to take action.

The routine nature of preparing and serving meals, purchasing foods, and planning menus provides opportunities to incorporate purchasing changes, operational changes, behavioral changes, and different food choices into dining operations. Dining may find that environmental issues can be incorporated directly into routine operations. For example, environmental inspections to ensure that recycling is carried out properly, doors are closed, and lights and equipment are off when not needed could become part of routine inventory or sanitation inspections. Proper times to turn on equipment to warm it and the actual ingredients used can be included as part of recipes and meal preparation protocols. Menus can feature organic, natural, vegetarian, local, and other lower-impact foods. These changes are most successful when they are incorporated into the culture and standard practices of the everyday operations and when staff are involved in developing the strategies that they themselves must carry out.

# 6

# Academic Departments, Administrative Offices, and Classrooms

The activities and choices made within offices and classrooms on college and university campuses offer many opportunities for environmental actions. These actions are symbolic of environmental commitment but also significant in that they influence vendors, waste generation, and energy use. University departments that pledge to reduce the waste generated from their activities by undertaking environmental stewardship actions can teach by example, reduce wastes, and reduce costs to the department and the university. Departments can strive to reduce and reuse paper and other materials by setting up new systems and changing expectations. Purchasing decisions in offices and classroom as well as behavioral choices also have great potential for tangible waste reduction and energy-saving results.

In offices and classrooms, the reduction of waste is the major thrust of environmental efforts, just as it is in buildings and grounds, purchasing, and dining services. Here, though, the waste reduction actions rely less on the installation of new technology or university-wide policies and more on routine, daily, and personal action. For this reason, office and classroom actions can involve many members of the university who feel that they have important and personal roles in the campus environmental movement. As elsewhere, these personal actions are most effective when they are backed by institutional policies and commitment, but office staff, faculty, and students need not wait for institutional policy in order to act.

Many universities characterize their environmental action only by their teaching and research, ignoring the wastes generated, chemicals used, and energy wasted in these educational pursuits. In addition to the choices about how to run an office or classroom, the topics addressed in the classroom itself have the potential to raise and address many examples of environmental action and protection. For example, a German class may want to study the aggressive German recycling laws that require manufacturers to take back used products, or an electrical engineering class may experiment with technologies to reduce the demand for electricity. This combination of hands-on campus-based actions with existing curriculum is effective for learning.

University offices also make other decisions with environmental consequences. Office equipment—computers, copy machines, and fax machines—collectively use large amounts of electricity. Departments' kitchen refrigerators and other appliances, water coolers, and restrooms have environmental consequences associated with energy and water use. The transportation of goods, students, and faculty on field trips and to research conferences and airports has an impact on air quality and climate conditions. Commuting and parking have impacts on the local and global environments. Many departments, from chemistry to the arts, use hazardous chemicals that require proper handling, storage, and disposal (see chapter 7).

### The Role of a University Committee

A university-wide committee can be instrumental in sharing and supporting environmental stewardship ideas among departments. Efforts that have been successful in one department can be promoted so that others can benefit from the experience. In addition, the committee can be instrumental in conveying new policies (e.g., purchasing policies) to departments or demonstrating new products. A committee can also provide a forum for understanding and cooperation between academic departments and operations departments, which generally are not administratively connected to one another. This forum can allow building, policy, and student issues to be explored collectively.

## Reducing Paper

The use of paper is one of the most visible uses of resources and generation of waste (although energy used in running the building, using equipment, and transportation may have more significant environmental impacts). Classrooms and courses are major paper users. Tufts makes more than 2,000 copies per student each year—and that excludes large copying jobs that are sent off campus. In turn, students relate their learning by taking tests on paper and writing papers.

In this country, paper and paper products, including copy paper, notebooks, books, magazines, paper towels, cardboard, and boxboard, make up about 40 percent of the solid waste stream, a percentage that is expected to reach nearly 50 percent by 2010.[1] In university offices, paper makes up nearly half of the entire waste stream.[2] In all cases, the largest component of this paper is writing and printing paper and printed material.

Paper is an integral part of student records, faculty research, and student learning. To process all of this paper, we use electric machines: computers, copy machines, printers, and fax machines. The ease of printing and duplicating documents makes waste easy and source reduction difficult. In addition, we have come to expect error-free letters and file documents printed on a single side of lily-white paper. For some staff, the production of error-free documents is a measure of their professional performance. This need for visible perfection, and the process of printing all draft documents, can be very wasteful.

The prevalence of paper means that there are countless behavioral changes and new technologies that provide opportunities to reduce the environmental impacts and the financial costs associated with paper use and disposal. Taking action to reduce paper use in the office and the classroom sends a clear message that environmental issues are important on the institutional and individual levels. At one university, a manager always sent single-side memos to his administration, rationalizing this decision by saying, "They don't read double sided there." Regardless of the accuracy of this perception, the administrators at this university had sent a message that wasting paper was acceptable.

Certainly recycling is important in offices and classrooms, but many university environmental efforts have focused exclusively on recycling. If university departments spent the same level of effort on reducing waste as they do on recycling, they would be more likely to address long-term environmental problems. Nonetheless, our experience at Tufts showed that waste reduction efforts at the department level are usually spearheaded by a single individual who tackles one decision at a time.

**Observe the Paper Stream**

In my own Tufts office, I find that waste reduction is largely out of my control, despite my best efforts to reuse paper, do double-sided copying, and use e-mail. Every day I receive stacks of mail, dozens of copies, and I feel as if I may soon be swimming in paper. When I look around, I realize I am not alone.

Most offices present plenty of opportunities to cut down on paper waste through reuse and reduction. The objective of observing the paper stream should be to identify specific solutions for waste reduction rather than gather lots of data (and use lots of paper!). In particular the audit should examine the following:

- How much paper in the recycling bin (or the trash) has a usable blank side
- The quantity and type of paper purchased
- The number of duplicate filing systems on the entire campus (for example, if one department loses a copy of a document, can it be retrieved from another department?)
- Recurring printer or copier problem, such as jams, that causes paper waste
- Draft documents reprinted in their entirety to fix minor errors
- Vendors whose packaging is wasteful
- Other areas where waste occurs

In addition, an examination of the recycling and trash bins will reveal the sources of paper from outside the university or department:

- Catalogs and junk mail sent to faculty or staff who are no longer with the department
- Multiple copies of the same publications
- Handouts and promotional material sent to the department in bulk for distribution to students

Anyone can take on the walk-through to assess the waste in an office. The most successful audit team will include generators of the material (usually office staff and students) and someone who understands the collection of the material (such as a building custodian). These assessments should occur at the end of the day, when barrels are full.

**Organize the Department to Act**

Once the information is gathered, it is time to develop and support specific waste reduction activities. The most successful efforts are outlined specifically and supported by a department's employees. Many proponents of waste reduction advocate the formation of a waste reduction task force with representation from throughout the office. Regular meetings of a group of faculty, staff, and students interested in waste reduction can help to foster new ideas and solve common problems. If department chairs can dedicate a few minutes of time in regular staff or department meetings to give waste reduction pointers and underscore the department's commitment to waste reduction, these strategies are likely to become part of a department's culture. Particularly at busy times of the year, faculty, staff, and students may find that it is easiest to bypass source reduction efforts, unless there is an institutional reminder that the effort is valuable.

Since paper use is so pervasive in department offices, there is a risk that the burden of the department's or university's waste reduction efforts will fall to the administrative support staff. To prevent burn-out of this important sector of the staff, waste reduction must become part of departmental culture, and conscious efforts should be made to avoid burdening a particular segment of the department staff. This is best accomplished by departments' ensuring that all of their members practice source reduction in their daily routines.

**Reduce and Reuse Paper in Offices**

It is probably not an overstatement to say that universities are addicted to paper—both reading it and using it to communicate ideas. Students at the University of Wisconsin at Madison estimated that fifty tons of paper were ending up in the recycling stream each school year without ever being read.[3] Reducing paper use requires that university personnel and students change expectations. For example, most of us expect documents on clean paper and use paper filing systems, but printing on scratch paper or storing documents on computer disk are equally effective (though less familiar). One department at Tufts embraced a goal of using all paper on both sides before recycling it.

Although it is often difficult to measure the results, waste reduction has many benefits to the university over recycling. Waste reduction has measurable financial savings resulting from the saved supplies and reduced waste hauling and disposal fees. Where possible, departments should try to quantify the results of their waste reduction efforts, by either tracking purchases or repeating the initial survey of wastes. Tracking departmental progress, even qualitatively, can provide important motivation and feedback that energizes future efforts. Table 6.1 outlines some common waste reduction strategies.

**The Effective Use of Computers** The effective use of computer technology can help reduce waste. Tufts found that it could reduce paper use by hundreds of pounds annually by simply reducing the type size it used to print monthly accounting reports. The change meant that the accounting department printed monthly reports on standard-sized paper ($8\frac{1}{2} \times 11$ inches) rather than larger ledger-size paper ($11 \times 17$ inches). Changing margins and fonts for file copies of reports or storing material on floppy disks reduces waste.

E-mail and electronic access to records are effective tools for reducing paper use. Many universities have systems for sending messages electronically to the entire campus by having a central person post the message on e-mail. This same person provides mailing labels for staff and faculty who do not have access to e-mail. The department initiating the message can send a paper copy to those individuals alone. At Tufts this change is estimated to have saved over 200,000 mailings each year.

Storing documents (letters, notices, student files) on disk rather than on paper can also reduce waste. Much like a file cabinet, a well-organized set of computer directories, clear file names, and regular backing up are essential to making computer filing systems work well.

**Double-Sided Copying and Printing** Double-sided copying and printing should always be the default method of printing and copying. Faculty should have large jobs, such as readers for courses, double-sided and evaluate each page to be sure that it is necessary.

Lack of knowledge about how to make double-sided copies is often the only barrier to full use of this feature (if it is available on the machine). Training and instructional signs can help overcome this problem. In fact, many small jobs can easily be copied on both sides by running the paper through the feeder twice. Fear of paper jams, especially when the need for copies is critical, is another legitimate deterrent to double-sided copying.

Since most universities lease large numbers of machines or contract for their maintenance, office personnel should demand that the machine operate dependably. Copy machines with double-sided capability generally are larger than those that can copy only single-sided pages and generally use more electricity. If an office is not using all of the features of its machine (especially because they do not work), it may be most efficient to trade it in for a simpler and more efficient model.

**Optimizing Existing Systems** Many universities have antiquated record handling systems. Computer science students are ready resources for consolidating information, making systems efficient, and reducing the waste in existing systems. For example, university mailing lists include university students, alumni, and employees. If a person falls into all three categories, she or he still needs only one copy of the annual report or a letter from the president. Careful database management can dramatically reduce costs and waste, and students are a ready resource for this purpose.

**Sharing Documents** Perhaps the most obvious but also the most difficult method of paper reduction is to share documents and publications. A distribution list of names of people who need to see a document is attached to the front of the document; each person checks his or her name off after reading it and passes it to the next person. The problem comes when one or two people keep the memo or magazine for weeks on end, but peer pressure can usually solve this. Leadership from the department chair or other prominent department figures can also be valuable in making this strategy effective.

**Reusing Paper** The results of the observation phase can point out many opportunities to reduce paper use. The most powerful of these opportunities will require changes in procedures, behavior, and expectations. For

**Table 6.1**
Source reduction strategies for offices

| Material | Source reduction strategy | | | |
|---|---|---|---|---|
| | Reduce | Reuse | Recycle | Other |
| Paper | | | | |
| White paper | E-mail<br>Routing slips<br>Double-sided copying<br>Fax cover sheet used only when needed; small fax sticky notes used<br>Preview documents before printing | Precycle by using paper on both sides; at Tufts and Mount Holyoke, old paper is made into precycled note pads<br>Print draft documents on reused paper | White paper is recycled at countless colleges and universities (Dartmouth's system has been heralded as a model)<br>The purchase of recycled paper is essential to closing the recycling loop; Northeastern buys paper from the same mill that it sells its waste paper to | The University of Vermont's recycling office annotates its reused envelopes with a stamp, "This envelope is reused to prevent waste"<br>SUNY at Stony Brook buys recycled paper for all of its uses |
| Mailings | Campus-wide mailings banned (Wesleyan, Franklin and Marshall)<br>E-mail used for mailings (Tufts and Dartmouth)<br>Self-mailing memos and written material<br>Two-way envelopes<br>Large mailings need prior approval (Wesleyan) | Reuse old envelopes | Make sure that mailing materials are recyclable in the university system<br>Include a reminder on the mailing: "please recycle" | Top university management should be clear that it encourages "source-reduced" mailings<br>Tag reduction and recycling messages onto other mailings, for example: "This mailing has been printed on both sides to reduce waste and printed on recycled paper" |

| | | | | |
|---|---|---|---|---|
| Computer paper | Computerized purchase order system (MIT)<br>Compress type to allow change from wide carriage to standard paper (Tufts saved about 2.5 tons of paper a year)<br>Financial information stored on microfiche (Tufts)<br>On-screen report viewing | Precycle computer paper by printing drafts on the back side of used paper | Computer paper is very recyclable, and its collection is common<br>Recycled computer paper is available although costly<br>Dot matrix printers make paper less toxic to recycle and use less energy when printing | |
| Bulletin boards | Use electronic bulletin boards or central newsletter<br>Limit to posting only one flyer per event (Wesleyan) with a basket holding additional copies | Reuse back sides of posters | | |
| Telephone books | Distribute one book per phone rather than one for each employee (University of Wisconsin)<br>Share phone books<br>Phone books printed on smaller pages with smaller type | Use old phone books for backup or in less frequently used locations | Phone book recycling at Brown captured 75 percent of its books at once<br>Phone books can be included with some mixed paper recycling (Tufts) | Promote e-mail by including e-mail addresses |

**Table 6.1 continued**

| | Source reduction strategy | | | |
|---|---|---|---|---|
| Material | Reduce | Reuse | Recycle | Other |
| Newspapers and other publications | Share subscriptions<br>Request that unwanted publications stop delivery | Newspaper is successfully used as animal bedding in some states (New York, Maine) | Newspapers are widely recycled | |
| Paper towels | Rolls of paper towels tend to use less materials than three-fold paper towels | Asbury Park Press in New Jersey saved $35,000 by switching to reusable cloth cleaning rags | Paper towels would be an appropriate bulking agent for a compost mixture<br>Paper towels are widely available with high quantities of postconsumer recycled fiber | |
| Packaging | Buy particular items in bulk (e.g., cereal)<br>Buy concentrates and mix on-site (e.g., cleaners) | Request that office furniture be shipped in blankets rather than boxes (Steel Case)<br>Return boxes, bread | Corrugated cardboard is recyclable<br>Use bidding process to specify the recycled content of cardboard boxes | Many schools have taken steps to eliminate, reduce, or reuse foam packing materials |

| | | | |
|---|---|---|---|
| | Use soap and shampoo dispensers to avoid individual bars (Park Plaza Hotel)<br>Request that large shipments come on pallets rather than in individually wrapped boxes (University of Wisconsin)<br>Bookstore asks if bag is needed; reusable bags available | trays, etc. to vendor for reuse<br>Reuse 5-gallon pickle buckets as recycling or trash bins (Tufts, Allegheny)<br>Reuse or return foam packing peanuts<br>Reuse of boxes by students at the end of the term | |
| Equipment | | Surplus equipment exchange (University of Iowa has a computerized list and estimates savings of $2 million)<br>Laser printer cartridges can be refilled<br>Printer cartridges can be reinked (University of Vermont) | Metals recycling is common |

example, reusing paper that is printed on only one side requires that the paper be collected and reinserted into the printer or copier. To be successful, it must be acceptable to receive memos, file documents, and drafts printed on reused paper. When it is difficult to identify the active side from the used side, a single line through the precycled side can help the reader ascertain which side to read.

Printers and copy machines with two trays can have one filled with reused paper and the other with blank paper. The default tray prints or copies on used paper. As an example, all internal drafts for this book were printed on reused paper. Similarly, an entrepreneur starting a company in Massachusetts used nearly five cases of paper over his two-year start-up; only two reams (1,000 sheets) were new paper, and the rest came from a large university office at Tufts that would have otherwise thrown the paper in the recycling bin. The savings to this one-person company was nearly $200. Universities count pennies in the same way that a fledgling company does, and multiplied over many departments, the savings can mount.

Offices can also reuse paper in scratch pads and telephone message pads. Some offices at Tufts collect paper for reuse in boxes. Office residents are taught to place paper flat in the box with the blank side up. The university printing shop cuts the paper into one-quarter or one-half size sheets, adds a cardboard backing, and makes a "Precycle" pad. This can be done at almost any printing shop for pennies per pad.

Reusing envelopes also reduces waste and cost, especially for intracampus mail. Mailing labels can cover old addresses when using envelopes to go off campus. The recycling office at the University of Vermont stamps reused envelopes with an educational message, "Reused to Reduce Waste."

### Reduce and Reuse Paper in the Classroom

The same principles of reduction and reuse apply to student class work. Students can be very creative if they believe that their creativity will be tolerated. In contrast to traditional term papers—double-spaced and single-sided—faculty can encourage students to reuse paper, even for final drafts, printed single-spaced or with one-and-a-half-line spacing to save paper. In response to student pressure, the Tufts graduate school now allows graduate student theses to be double-sided (although still double spaced), thereby reducing the volume of paper for these documents by as much as 50 percent.

Further paper reduction could result if students delivered their papers electronically, a change that would require faculty to grade and comment on papers differently. In fact, some faculty say that they make more in-depth comments on the electronic versions of student writing than they had on the paper versions. Faculty are particularly reluctant to change, but encouragement from students and administrators may help them to experiment with techniques that reduce waste.

Student readers, handouts, and materials for students working in groups can all use less paper and generate paper waste. Some material can be provided electronically. Placing copies of readings on library reserve or providing teams of two students with a single copy of the readings can also help decrease paper use, provided students do not make additional copies on their own. Students working in groups often feel an urge to have copies of all documents for each team member. A central mailbox system or a system of reading material and passing it along can consolidate research material and reduce waste. Faculty members should lead by example by limiting copies, putting material on reserve, and encouraging students to reduce waste from their activities in and out of class.

**Reduce Paper Use in Computer Centers**

Computer centers generate vast quantities of waste paper as students print drafts of term papers, computer programs, and e-mail messages. Printing all but final versions of these documents on precycled paper can reduce costs and waste generation. Collection of the paper on a voluntary basis can itself inform students that their work will be used by others as scrap paper. Computer center users can be informed that paper will be reused and that confidential documents should be torn in half before recycling. The goal of the computer center should be to have all paper printed on two sides before it is recycled.

**Reduce Incoming Paper**

Much of the paper waste found in the assessment of department waste will come from outside the department or even off campus. Reducing paper that comes into a department can be time-consuming. The most effective strategy is to identify its source and send postcards, usually with the mailing label, requesting that the mail be stopped. The University of Oregon's recycling coordinator provides postcards free of charge to departments wanting to reduce excess mailings. Putting names on the Direct Marketing Association's list to stop junk mail can also be effective. Mail the name and address to be removed to Direct Marketing Association, Mail Preference Service, P.O. Box 9008, Farmingdale NY 11735-9008 (212/768-7277).

**Avoid Being Part of the Waste Circle**

On some campuses, university administrators sell student and staff mailing lists to mailing houses that use the lists for bulk (junk mailings). Universities should prohibit this practice and be careful to avoid any excess mailings of their own.

**Provide Feedback on Waste and Waste Reduction Progress**

Per sheet, paper is cheap. One big reason that we waste so much of it is that few of us see the costs of the paper we use—its cost to our department, the cost of its disposal, or the cost of its production on the people who live near the paper plant. Although it is neither practical nor efficient to charge for each individual sheet of paper, departments should find ways to make department use of paper and other supplies evident and meaningful. Signs and charts tracking paper purchased (as it decreases) and follow-up examination of wastebaskets and recycling bins will help point out progress. Departments should take care to measure real waste reduction (paper per student or paper per faculty member) rather than something directly linked to the natural growing and shrinking of a department.

**Recycle the Remaining Paper**

Despite its best efforts, every office will have waste paper, and the remaining paper should be recycled in the university's recycling program. Chapter 3 describes some key elements of setting up a recycling program. However, successful recycling in the office requires sorting on an individual basis.

Training of department staff reinforced by clear and well-placed signs is important to the success of a recycling program. Most offices find that assigning recycling responsibilities to one individual can help identify problems and provide information to new teaching assistants, faculty, or office staff. Of course, the responsibility for recycling in each office falls to individuals themselves. Diligent monitoring and repeated training improve recycling success in an office. The Environmental Task Force at Church and Dwight Co., makers of Arm and Hammer products, has given out recycling tickets to praise or to point out where progress is needed. Members of student environmental groups may be likely candidates to serve as "recycling police," although this type of oversight is likely to be resisted, especially in a university community. "Policing" may be effective in some settings, but it is likely to turn people off to both recycling and other environmental efforts in the long run.

When departments purchase materials, particularly paper, they should consider if the material is recyclable in the campus program. For example, a college that recycles only white paper should discourage academic and administrative departments from purchasing yellow-lined legal pads since the colored paper is not as valuable in its recycling programs. And faculty members can encourage students to use recyclable paper for notes, lab books, and graph paper.

**Buy Recycled and Unbleached Paper**

Using recycled paper is important for ensuring markets for the paper collected by campus recycling programs. The production of recycled paper uses less raw material, and generates less waste than the manufacture of virgin paper. Using unbleached paper, or paper bleached without chlorine, has additional benefits in reduced emissions of dioxin, a suspected carcinogen.

Recycled paper is available for almost every job. Sometimes it costs about the same as virgin paper, but it can be more expensive, although this is changing. Universities that buy paper in large volume may find that recycled paper, bought in quantity, is price competitive. In addition, the source reduction efforts of a department can more than offset the incremental costs of recycled paper. Ideally, all university paper should have recycled content. If the institution is unwilling, individual departments can make the commitment on their own.

There is an intellectual debate over the appropriate target amounts of postconsumer waste paper (paper collected from commercial and residential sources) versus preconsumer paper (milling scraps) that should be in recycled paper. Those who advocate that recycled paper should have the highest possible postconsumer content seek to fully support recycling. Others argue that recycling of preconsumer waste should be equally supported. William Rathje, an archeologist famous for studying garbage, advises that anything over 10 percent postconsumer waste is worthwhile.[4] EPA's current standard for government paper is 20 percent postconsumer waste. Although increasing the use of postconsumer paper helps support the recycling efforts on campuses (if only indirectly), it is important that the details of this debate not overshadow an environmental commitment and preclude a decision to use recycled paper at all.

Unbleached paper is less readily available, but several major suppliers are now carrying lower-quality unbleached recycled paper (for example, Hammermill's Unity DP paper made from recycled newspapers and magazines) at prices that are lower than bleached white recycled paper and some bleached white virgin paper customarily used in most offices. In places where cost is always the overriding argument, consider switching to this kind of unbleached paper. It works just as well for most office applications despite the lower quality. In some applications, especially when copying from an unbleached copy, the quality is poor. Nonetheless, unbleached paper is excellent for single-distribution memorandums and file copies. Paper bleached with nonchlorine bleach is also preferable to standard paper stock. Departments can ask their university purchasing departments to make it available.

Tree-free paper made from hemp, straw, and kenaf is also available. These alternative types of paper are more expensive than wood-based paper but may be more environmentally sound since these materials can be grown more productively (more fiber per acre) than wood. The University of Oregon's print shop is experimenting with offering these tree-free papers.[5]

Students can use recycled paper for papers, homework, and notes. Faculty can encourage this use by providing students with information about sources of recycled paper. The additional cost of recycled paper is not more than about fifteen cents for a twenty-five-page paper, and so requiring the use of recycled paper is not unreasonable.

## Office Equipment and Supplies

Offices purchase a range of supplies and services from pens and pencils, to desks and chairs, to electronic equipment. Each of these has environmental consequences resulting from their use, production, packaging, and disposal, so the principles of waste reduction can often help academic departments make sound choices that are best for the environment.

### Buy What Is Needed

There are many examples around almost any academic office where supplies are purchased in excess, subscriptions are renewed to publications that are never read, and duplicate equipment is purchased because people are unwilling to share. Buying only what is needed is a simple and cost-effective strategy for environmental stewardship.

Like the federal government, many universities budget their funds to departments, programs, or schools on a "use it or lose it" system: any money not spent in the current fiscal year is lost. Worse, departments may receive lower budgets if they failed to spend all of their budget in the preceding year. Universities that are willing to develop alternative policies will find that they can save money and avoid the overpurchasing of supplies. Departments can avoid the generation of waste with end-of-year spending by purchasing only needed or long-lasting educational materials or equipment. They can also reduce waste by purchasing high-quality, durable items, especially those that get hard wear. For example, desks and chairs get heavy use (and abuse) in university offices, so durability and quality will pay for themselves over time. Durable items are a good example of life-cycle benefits—that is, the actual cost of an item over its life, from manufacturer to disposal.

### Buy Energy Efficient Equipment

Once energy-efficient computers and office equipment products are available through university purchasing departments, it is up to departments and individuals to select them. University departments can promote the use of the most efficient equipment by restricting equipment to that with energy-efficient designation, such as computers with EPA's Energy Star designation for efficiency, and by promoting the benefits of energy efficiency and products that achieve that goal. Since Energy Star computers are available for almost every application, there are few downsides to this decision (see chapter 4).

### Use Technology to Facilitate Equipment Sharing

Since the late 1980s, nearly every university student, faculty member, and staff assistant has received access to a computer and printer, often right on their desks. Networking software and switch-box technologies facilitate sharing printer and other computing equipment such as modems, scanners, and disk drives. Sharing equipment is cost-effective in terms of both the equipment purchase price and the energy used by the equipment. University departments of computer services can promote and assist with equipment sharing.

### Turn Equipment and Lights Off When Not in Use

The easiest and most effective energy-saving action is to turn off lights and equipment when they are not needed. Contrary to myths that say that turning off lights wastes energy, lights should be turned off anytime anyone leaves a room.[6] Although switching lights on and off frequently will shorten their life, the number of hours of useful service is greater since the lights are using electricity only when they are on. Leaving the lights on all the time may provide more "burn hours" from the light bulb (known by lighting professionals as a lamp), but a portion (often a large portion) of these hours of use may be when the room is empty.

Computers and copy machines should be turned off at night and on weekends. Turning them off, especially computers, during meetings, meals, and other extended periods of inactivity is also prudent. If the staff or an office is prone to forgetting to switch off their office equipment, an automatic timer can be installed to turn it off at day's end.

### Find Ways to Reuse Old and Broken Equipment

Broken and outdated equipment can be reused and repaired. Often this equipment is a valuable asset for local public schools or for colleges and universities overseas. Old equipment that too often clutters the back halls and attics of the universities can be put back into productive service by finding a partner who can get it to someone who will use it. State commerce department or international aid offices can assist departments in finding these partners. For example, Dave Harrington at the Tufts New England Medical Center has been collecting used and broken medical and electronic equipment for the American Medical Resources Foundation (AMRF) in Brockton, Massachusetts. AMRF repairs microscopes, oscilloscopes, and diagnostic and other equipment and donates it to medical and teaching institutions overseas. In its five years of activity, the AMRF has donated supplies that would otherwise have been thrown away, to eighty-four hospitals in thirty-six countries—among them, Nicaragua, Ecuador, India, Russia, Vietnam, Paraguay, and Brazil.

Used textbooks, slightly out-of-date collections, and recently updated editions can also get a second life. Following the example set by the University of New England, students and staff at Tufts worked with the Tufts Bookstore to collect more than 400 used textbooks from students departing at year end. These books were shipped to the Federal University of Matogrosso in Brazil, where they are a much appreciated addition to the university library. The International Book Project also accepts textbooks under ten years old and journals under five years old for shipment to overseas countries. Most of their books are used in primary and secondary schools, universities, libraries, churches, hospitals, or Peace Corps–affiliated libraries.[7]

Even on a local scale, there are often underground markets for materials. For example, a Tufts staff member whose husband is starting a small business takes the Styrofoam packing peanuts from several departments home, where they are reused as packing material for this small business. Most Mailboxes Etc. stores accept packing peanuts, which they re-sell. And at the end of each semester, the campus bookstore and dining halls can provide boxes to students for shipping and storing belongings. Entrepreneurial students, employees, and community environmentalists can find similar arrangements with local organizations and develop relationships that are mutually beneficial.

## Conferences, Meetings, and Special Events

Conferences, meetings and other special events held on or off campus are often important point sources of waste and intense resource users. When these are environmental events, there is pressure to prove that change begins at home.

Events that keep wastes at a minimum set an example to those who attend the events or may use conference or meeting facilities in the future. Most university events run on tight budgets, so waste reduction and environmental stewardship efforts should help organizers meet fiscal and environmental goals. Waste reduction efforts can offset the small premium of using recycled products or collecting and recycling any wastes or specialty items. Recylability should be considered when selecting any paper or single-use items that cannot be avoided.

### Take Conscious Waste Reduction Steps

Waste reduction is one of the easiest things to undertake in conferences or meetings regardless of their size and location. The following list provides some suggestions:

- Minimize paper used in initial mailings, invitations, or promotional materials.
- Provide preconference proceedings on disk or on-line.
- Print all copies on both sides of the paper.
- Post announcements, agendas, and other information that does not have lasting benefit rather than provide copies to all participants.
- Use reusable name tags, and provide collection bins at the end of the meeting.
- Use reusable dishes and serve bulk drinks rather than individual cans.
- Recycle all wastes that remain, or donate leftover posters and other supplies to local schools for reuse.
- Use recycled paper printed with soybean inks.
- Do not provide folders unless participants need them.
- Do not feel compelled to provide conference give-aways (cups, magnets, pens, etc.).

### Ask for Cooperation from Conference Organizers or Facilities

When a conference or meeting is being run by a conference organizer or is held over several days at a hotel or other facility, it is reasonable to expect that environmental objectives will be carried out by these service providers. Departments or organizations hosting the conference can convey their expectations to the conference organizer, hotel, or meal provider ahead of time. Waste reduction steps in the facility might include reusable dishes, reusable table cloths, and recycling for paper. Conference hosts can even consider the environmental track record in selecting a conference facility.

Conference organizers can provide a number of services to attendees that are a benefit to them and to the environment. For example, they can offer a carpooling service to match rides for people coming from out of town or provide a shuttle service from the airport in order to reduce the number of single-occupancy trips. Organic food and vegetarian meals are environmentally friendly and often healthier alternatives to standard cuisine.

## Transportation

Every day faculty, staff, and students commute to college and university campuses, many of them alone.[8] Tufts University estimates that the average daily commute of staff and faculty is twenty-seven miles and that 75 percent of these commuters travel alone in private automobiles. As a result, vast amounts of air pollutants—hydrocarbons, carbon monoxide, nitrogen oxide, and sulfur dioxide—pour into the atmosphere, contributing to the diminished air quality of the region. In Massachusetts, the Department of Environmental Protection's mobile sources (cars and trucks) generate 48 percent of the state's volatile organic compounds, 47 percent of its nitrogen oxides, and 70 percent of the carbon monoxide. In addition, almost one-third of the state's carbon dioxide, the principal human-produced greenhouse gas, comes from cars and trucks.

There are four major reasons to increase the efficiency of the existing transportation network of students and staff at a university:

1. A transportation efficiency program is likely to be dramatically more cost-effective than the construction of parking spaces.
2. The Clean Air Act of 1990 requires these transportation reduction strategies of any university in an area that does not meet air quality standards.
3. Improving transportation efficiency is consistent with a green university's commitment to environmental stewardship.
4. The proposed strategies benefit the surrounding communities by decreasing vehicular traffic and congestion.

Cornell University undertook an aggressive transportation demand management program in 1991 and in a single year reduced the number of faculty and staff vehicles brought to campus each day by 26 percent. The program included an awareness campaign, alternatives to parking through mass transit and ride-sharing programs, and higher parking fees. The program resulted in an estimated net annual saving of 417,000 gallons of fuel and the associated 6.7 million pounds of carbon dioxide, 651,000 pounds of carbon monoxide, 34,000 pounds of nitrous oxides, and 59,000 of hydrocarbons. In addition the program saved over $2.7 million in four years by avoiding the cost of constructing new parking spaces![9]

**Take the Lead on Complying with New Requirements**

In the 1990 Clean Air Act Amendments, Congress revised the guidelines for compliance with EPA air quality standards. The amended law ranks levels of air pollution in American cities according to their severity and sets deadlines for compliance with air quality standards. If air pollution in metropolitan areas is ranked serious with respect to ozone (a by-product pollutant of vehicle emissions), that region has until 1999 to meet air quality standards. To meet the federal air quality standard by 1999, officials in many states are considering strategies for reducing auto emissions. By preparing now, a university can place itself at the forefront of compliance with the Clean Air Act and make the future transition smoothly. Regulations that affect universities include employer trip reduction requirements. These will require companies of a certain size (usually employing 100 or more people) to reduce the number of employees commuting alone in their cars each day. Some municipalities, such as Cambridge, Massachusetts, are considering such a requirement as well. The Energy Policy Act will also require organizations with centrally fueled fleets of vehicles to use clean-burning fuels in some portion of the fleet.

**Promote Ride Sharing**

Sharing rides to and from work is one of the most effective parking-demand management and environmental conservation strategies. Ride-sharing systems work effectively in many corporations and hospitals in the country, especially in southern California. The ride-sharing programs:

• Create carpools and vanpools and match riders by computerization and publicity.

• Guarantee emergency rides home (e.g., for a sick child or home emergency) to employees who share rides.

• Create incentives with preferential parking spaces and lower or free parking rates to those who ride-share.

### Take Steps to Increase the Use of Public Transit

Many universities are relatively well served by public transportation—buses, university buses, or subways. Some simple steps can facilitate use of the public transit by university employees and students:

• Participate in a transit pass subsidy program for students and employees. Such a program offers monthly passes to employees at a reduced rate. The discount may be paid for by the university or by the transit authority. Cornell provides free transit passes to those who do not purchase parking passes.

• Sell transit tokens and provide exact change for buses in departmental offices.

• Work with the local transit authority to have the schedules and bus routes meet university needs.

• Improve awareness of schedules by publicizing them in the central locations as well as the central office of research centers and academic departments. Post and weather-protect schedules (inside waterproof bulletin boards) at the stops along the campus.

• Offer contract or demand-driven shuttle services whereby a bus route is created to meet the needs of employees or students who live near each other in a specific neighborhood or region.

### Facilitate Bicycle Use

Promote bicycling by placing adequate bicycle racks around campus. Investigate the campus to identify where additional bike racks are needed; bikes locked to railings and sign-posts are clear indications of a need for additional racks. Regular cyclists can also easily provide this information. Locate bike racks close to major buildings and other destinations, and ensure that the racks provide security for well-locked bikes. Bike racks should be located in well-lit and frequently patrolled areas.

The University of Illinois' Urbana-Champaign campus has established a sophisticated network of bike lanes and bike parking facilities. The bike

**Table 6.2**
Pollution and other costs of commuting 10 miles

| Mode of commute | Initial cost | Financial cost to commute 10 miles each way | Environmental costs of the commute | Other issues |
|---|---|---|---|---|
| Bike | Cost of bike and helmet: \$150–\$350 and up | Minimal | | Safety, time, need to change clothes, exercise, foul weather |
| Public transit | None | \$1–\$3 per day | Low per person | Schedule and route may be restrictive |
| Drive alone | Car purchase and insurance | \$5.40 per day | | Most convenient |
| Drive with one other person | Car purchase and insurance | \$2.70 per day | | Schedule may be restrictive |

lanes are removed from sidewalks and roadways and are well marked.[10] This aggressive approach to encouraging bike use is most effective.

### Conduct a Public Education Campaign

Studies show that changes in consumer habits can be achieved through public education.[11] A public education campaign on commuting at a university can inform people about the social and personal cost of driving and of the availability of alternatives.

Most universities have an individual or department responsible for traffic and parking. Often it is a part of the public safety (campus police) or human resources department. A transportation coordinator can create a traffic and parking regulations booklet or Internet site that is distributed or otherwise made available to faculty, staff, students, and visitors. The booklet might include information about the pollution and financial costs, per person, of the daily commute under different modes—alone in a car, in a carpool, by bus or other public transportation, and on a bicycle (see table 6.2)—and about how to get to the campus: driving directions,

directions by public transit, and a bus schedule. In addition, the coordinator can promote alternative commuting habits with new employees by providing them with bus and shuttle schedules, offering to enter their name in the ride-share database, and providing safe lock-up areas for bikes. Further, a well-publicized and centrally located ride board can help students, faculty, and staff to find others who need a ride or can offer a ride on a one-time or regular basis.

### Increase Parking Rates

An increase in parking rates will be the most difficult strategy to implement at almost any university unless price changes are offset by adding commuting benefits. Cornell's transporation plan included increasing parking fees in order to deter single-occupant vehicle use and decreased fees to encourage ride sharing.[12] Parking rates should reflect the cost to construct parking and the cost of maintenance, lighting, police enforcement, plowing, and other routine activities. These real parking costs should be distributed across parking rates. Parking rates might be prorated on the basis of employee salaries to avoid overburdening lower-paid employees with higher parking rates. True parking costs might be about 0.5 percent of an employee's salary. Smaller cars could also be rewarded with lower parking rates in order to encourage those who drive energy-efficient vehicles.

### Encourage Ride Sharing and Public Transit When Conducting University Business

The environmental impacts that result from commuting to and from campus may be very hard to control except by providing incentives and rewards. Travel on university business, both locally and to other cities, is something over which the university has some jurisdiction. In a large university, it is likely that there are numerous trips to and from the airport on any given day. Sometimes several taxicabs arrive at the university at the same time to take people from several departments to another campus, a frequently visited off-site location, or the local airport. Coordinating the use of taxis through a single cab company or a central dispatcher can reduce the miles traveled and the costs of duplicate transportation. Universities with large campuses or several locations should coordinate rides to meetings at a satellite campus when it is not uncommon for several administrators or faculty to each drive private cars to the same meeting.

In the same way that many universities, government agencies, and the private sector limit per diem expenses for meals and lodging, the green university might reimburse traveling faculty, administrators, and staff only for the cost of travel by public transit in cities that are well served by subways, buses, and rail. Employees who request additional reimbursement would have to justify their expense.

## Curriculum

When universities set out to promote environmental commitment, they usually focus on their academic programs by developing or strengthening specialties in environmental studies, environmental engineering, environmental health, and ecology rather than on their efforts to install energy-efficient heating systems or state-of-the-art composting. Focusing environmental education in a single department or several academic specialties that focus on areas such as natural resources, sciences, health, and engineering ensures that we educate specialists who can monitor environmental conditions, remediate contaminated sites, invent and install pollution control technology, or litigate environmental violations. This approach usually may not provide complete knowledge or understanding of environmental problems and their solutions since environmental issues are complex and cannot be understood through a single discipline or department.[13] All of our actions have environmental consequences, and many of them are not apparent to us. Anthony Cortese argues, "Because all members of society consume resources and produce pollution and waste, it is essential that all of us understand the importance of the environment to our existence and quality of life and that we have the knowledge, tools, and sense of responsibility to carry out our daily lives and professions in ways that minimize our impact on the environment. That is, we need an environmentally literate and responsible citizenry."[14]

To achieve this goal, Cortese proposes introducing environmental concepts in all disciplines.[15] In this way we can demonstrate the inherent connection of history, psychology, chemistry, and romance languages to the care and stewardship of the planet. Cortese's ideas have been embraced by a number of university faculty and universities themselves. There is as well a growing field of eco-psychology that acknowledges that the solutions to environmental problems must extend beyond technology and science to include an understanding of their nonphysical or behavioral aspects.

Growing from Cortese's idea and Tufts President Jean Mayer's goal of graduating environmentally literate citizens from every discipline at Tufts (including the medical, dental, and veterinary students), the Tufts Environmental Literacy Institute sought to teach faculty from Tufts and elsewhere about general environmental issues with a one- or two-week institute. Faculty from departments throughout the university, including drama, Spanish, English, and engineering, work together to understand the issues and relate them to their teaching. Most of the faculty are expected to revise their course offerings to reflect the summer institute's program. Evaluation of the success of the Tufts environmental literacy effort has been difficult, in part because it is hard to agree on what constitutes this knowledge. Nonetheless, the program has demonstrated that environmental issues can be taught as part of every discipline and that an interdisciplinary approach to thinking about the issues is beneficial and more accurately reflects most environmental problems. The University of Northern Iowa uses a different approach by requiring an environmental literacy course of all students.

David Orr argues for a more in-depth and sweeping change of education that begins with a commitment to life and its preservation, a motivating and energizing force that underlies education and research.[16] Orr believed that new ways of approaching education are needed for sustainability and ecological literacy. One aspect of Orr's proposal is to link a university's commitment to environmental education and educational reform with environmental stewardship in campus operations. This can be powerful for student learning and environmental change on a campus. One part of this linkage is for operations departments to capitalize on and respond to student energy and enthusiasm by using students as researchers, advocates, information sources, and sources of energy. A second part of this linkage is to use the university and its environmental decisions as teaching laboratories and hands-on case studies.[17]

**Link Student Learning with Campus-Based Action**

Students' academic projects, such as papers, theses, and group projects that are part of existing courses or independent study are excellent opportunities for linking student learning with campus-based action. This can occur as a result of student initative or under the direction of a faculty member. The University of Wisconsin has created a course that uses the campus as a laboratory. It involves a client-based project where the client is a member (usually staff) of the university community.[18]

Student projects in nearly any course can use the campus for a laboratory. For example, mechanical engineering students can evaluate the use of new energy-efficient motors in target locations in cooperation with the building and grounds department, a sociology student might find local foods for dining services, and a business major could evaluate the cost and benefits of a waste reduction program with the purchasing department.

Regardless of the type of project, its success hinges on the student or faculty member's investing sufficient time in assessing the feasibility of a topic and involving university staff at the outset. Feasible topics are those that can be covered in the time allowed, have issues that are germane to the course and teaching objectives, and are of interest to the university. For example, a student project to examine the use of reformulated paint on campus could be feasible and relevant if university staff paint university buildings; it would be less relevant if all painting is hired to outside contractors. Similarly, a project that describes wind-powered electric generators is of little interest to a university in a sheltered valley.

Feasibility is more likely if staff involvement begins in the project conceptualization, so that the final study meets their needs and interests. Students and faculty should solicit project topics directly from the university community and find employees to act as official contacts for each project. University staff in the buildings, grounds, purchasing, computer services, food services, and print shop departments are places to start finding project topics. Staff can be at almost any level, although finding someone with decision-making authority will help the project succeed. Because university decisions, such as those about paint selection, printer use, and athletic transportation, are complicated, in terms of both their environmental issues and practical considerations, it is important to limit student environmental projects to concrete and manageable small projects rather than to undertake the development of generalized recommendations that are merely concepts.

Developing project topics that meet university needs helps to motivate students. At Tufts, we found that the early involvement of members of the Tufts community was essential. Otherwise, university employees who were contacted by students were very likely to see the students as an intrusion rather than a resource. When the topics are of little interest to the university decision makers, students too become frustrated since the staff needed as sources of information seem uncooperative or unreliable.

### Use the University as a Laboratory for Faculty Research

Faculty can use the university as a laboratory for their research or to pilot-test new technology. Monitors to identify leaking underground storage tanks, techniques to improve boiler efficiency, and educational techniques to change behavior can each be tested on campus. Since the university staff are among those responsible for environmental action in university operations, they are potentially valuable sources of information. Faculty-staff collaborations can help keep research funds on campus and provide valuable information and technology to the university itself.

## Conclusions

The actions in offices and classrooms that support the university's educational and research missions can reflect a commitment to environmental stewardship and minimize the impacts. If made part of the everyday business of teaching and administering university business, these actions need not be burdensome and can send repeated and clear messages of commitment and action. Course work and faculty research can benefit from and contribute to efforts to green the campus. Combining these facets is an important part of beginning the transformations needed to reduce a university's environmental footprint and developing students, as well as university employees, who are environmentally literate.

# 7

# Laboratories, Research Facilities, and Studios Using Hazardous Materials

Hazardous materials—such as chemicals, chemical by-products, chemical handling supplies (e.g., gloves and spent bottles), paints, and solvents—can cause pollution and present risks to health, safety, and the environment. Flammables, corrosives, toxics, radioactive isotopes, and biohazards are used to teach chemistry, biology, chemical and civil engineering, physics, and geology. Laboratory research in these disciplines, as well as in biomedical and medical fields, also use and generate hazardous chemicals and hazardous materials. Physical hazards such as liquid carcinogens, sharps (used blades and needles), and lasers pose concerns. In addition, regulated and hazardous materials are used and waste is generated outside the sciences in places like photographic developing laboratories and art studios.

The use and disposal of chemicals have consequences for the health of the people who handle the material and those whose water, air, and land may be polluted by leaks, spills, and volatile emissions. The seriousness of these impacts has resulted in a growing number of laws and regulations that govern how hazardous materials are handled; universities were once exempt, but they are rarely so today. The financial and legal costs associated with complying with these regulations have also increased. Numerous reports must be filed with state and federal agencies on the use, transportation, and disposal of hazardous materials. Disposing of a few ounces of a material can cost under one hundred dollars if done correctly; cleaning up a spill of the same material can cost thousands of dollars. In addition, institutions are liable for violations of environmental regulations and may be liable for future health problems or environmental contamination resulting in the future from actions at the university today.

Under the law, civil and criminal responsibility may extend to the responsible personnel within the university, including the faculty, administration, and even the president. Thirty-five colleges, universities, and schools in Massachusetts alone are "potentially responsible parties" (PRPs) at Superfund sites, usually as a result of past waste disposal practices or the practices of their waste haulers. Under the legislation, these PRPs may be held financially liable for the entire site cleanup, regardless of the size of their contribution.

In industry and government, the focus of improved hazardous materials management has changed from concern for proper waste disposal to pollution prevention—preventing the generation of pollution in the first place, often by reducing or eliminating the use of certain materials altogether. Pollution prevention is becoming increasingly commonplace in industry, but academic laboratories and studios have been slow to embrace the concept despite the benefits and the real need to improve current university practices.

Academic laboratories differ from industrial processes in ways that may make pollution prevention difficult. Most important, university laboratories use small quantities of many different materials; in contrast, most industry deals with larger quantities of a few chemicals. At some universities, more than two thousand chemicals are frequently used, and often there are many containers of a single chemical.[1] Generally faculty decisions about laboratory curriculum are independent of the cost and payment of waste disposal or hazardous-spill cleanup costs, so the costs of the supplies or their disposal are not apparent or considered in the decision making. Furthermore, research projects are funded by outside agencies or foundations that are interested in results, not proper waste handling.

Meeting pollution prevention objectives nevertheless can complement the research objectives of a university and its faculty. This result requires incorporating a culture of reducing or avoiding wastes wherever possible and will save both money and time in the long run. In the green university, departments and faculty embrace pollution prevention in all courses, research, and even staffing, tenure, and promotion decisions. As the educators of future researchers, doctors, and other professionals, universities have an additional responsibility to embrace the principles of pollution prevention not only to comply with regulations and reduce costs and liabilities, but also to serve as a teacher of new methods and safe thinking.

## Meeting and Exceeding Legal Requirements

Numerous federal, state, and local laws govern the use of chemicals and the disposal of wastes, among them, the Resource Conservation and Recovery Act (RCRA), Superfund, the Clean Water Act, and the Clean Air Act. The Occupational Safety and Health Administration regulates worker safety and requires plans for safe chemical handling in laboratories, regulates the necessary safety equipment, and regulates the communication of risks to workers. Medical, infectious, and radioactive wastes are regulated as well.

The specific requirements that universities must meet are extensive, but there are a number of administrative actions that can help the institution to meet and exceed legal requirements, and avoid accidents and fines in the process.[2]

### Commit to Meeting or Exceeding Legal Requirements

Meeting the legal requirements for hazardous materials use and disposal requires full cooperation from students, staff, and faculty who work with these materials. Commitment from the administration to the principles of pollution prevention has demonstrated widespread benefit in terms of public relations, cost reduction, regulatory compliance, and worker productivity at corporations throughout the country.

### Hire an Environmental Health and Safety Staff

The regulatory nature and complexity of chemical and waste handling on a college or university campus requires that someone be responsible for overseeing compliance. In some places, departments handle their own waste disposal and permits, but most institutions find that it is efficient and cost-effective to have some staff time dedicated to these issues on a campus-wide or university-wide basis. In most institutions, this responsibility requires full-time staff, although in smaller institutions it is sometime designated on a part-time basis to people in the campus police department or the purchasing department. Environmental health and safety (EHS) presonnel should be knowledgeable in areas such as industrial hygiene, biohazards, radiation, chemical management, spill cleanup, and hazardous waste disposal. The EHS staff performs a range of activities, including training, spill cleanup, and laboratory auditing. Filing the required federal and state reports and tracking hazardous waste disposal manifests are essential and time-consuming parts of the job.

Funding for the EHS staff varies by university. Most universities consider funding the position or department from the general overhead budget.

**Form an Environmental Health and Safety Committee**

A committee charged with looking at the environmental health and safety aspects of research labs and studios using hazardous materials can have a positive effect on reducing and improving hazardous materials use. The committee can provide an important avenue of communication about new requirements and new techniques and a forum to evaluate and learn from accidents, spills, and new teaching methods. It is helpful in overcoming the autonomy with which most researchers and faculty operate, can provide support to research technicians or other staff when dealing with problems in their departments, and can encourage departments to include safety issues as regular agenda items in departmental meetings.

At Tufts, the committee was formed by EHS staff and the vice president of finance in order to address spiraling hazardous waste disposal costs. The committee is composed of a cross-section of individuals from different departments and different levels in the institution and includes the EHS staff, representation from all departments that use hazardous materials (including the art department), someone from human resources and buildings and grounds, and a police/safety officer. Other members might include legal counsel and a representative of the administration. The committee chair is usually more effective if he or she is a researcher or faculty member rather than an EHS staff member.

The committee usually facilitates communications and serves as an advisory body rather than an enforcement body. It reviews existing policies and procedures and recommends changes to improve the environmental impact of the institution. For example, the committee might discuss new

**Box 7.1**
Sample projects for an EHS committee

Information sharing, such as new regulations
New procedures, such as training for staff and students
New policy development, such as policies about chemical purchasing
Problem identification and solving
Disciplinary actions

regulatory requirements, ways to train staff and students about chemical handling, university laboratory and studio audits, and ways to share supplies. Box 7.1 shows some sample projects that might be undertaken by an EHS committee. One important benefit of an EHS committee is the peer pressure it can exert. At Tufts the committee is the only mechanism for disciplinary action of tenured faculty who make egregious safety mistakes. On occasions, these faculty have had to answer to this committee of their peers for their actions. Although no formal action results, the experience itself seems to be a punishment.

Chemical safety and handling can be sensitive issues for the university, and they are also complicated and heavily regulated. Thus it may be effective to have a university-wide environmental committee delegate these issues to a laboratory safety committee or a safety subgroup rather than to undertake them in a larger committee setting. A laboratory committee can use a university-wide environmental committee to endorse policies and oversee hazardous materials that are not covered in labs and studios. In addition, the university-wide committee can help ensure awareness of and support for campus policies about chemical handling and waste disposal.

### Clarify Responsibilities

The EHS staff is generally responsible for overseeing and helping the university meet its legal responsibilities for handling, using, and disposing of hazardous materials. Successful compliance requires individual responsibility and cooperation as well. Faculty, researchers, students, and artists tend to work independently, from both each other and from the administration, and there is little university oversight on the actual mechanics of the research or creative processes and their related chemical handling practices. Academic freedom and the status quo are often barriers to changing habits in the lab or studio. For this reason, the responsibilities of individual researchers, faculty members, research assistants, and teaching assistants for safe chemical handling, storage, and disposal should be clearly spelled out and regularly reinforced. This clarification of responsibilities can be undertaken by the EHS staff with the support of university administrators and the EHS committee.

Faculty and researchers commonly are unclear about university policies, legal requirements, and personal liability regarding the use and disposal of hazardous materials. University policy can establish clear guidelines and expectations for the use of these materials in order to ensure the safety of students, workers, and the environment, as well as to protect the institution from liability and litigation. Training can highlight the personal liability that each individual assumes (both legally and morally). The goal is for university personnel to understand that health, safety, and environmental protection are required by laws and as a condition of university employment. In turn, faculty can expect students to understand that responsible practices are requirements of their courses. Providing lab users with information about the cost of accident response, cleanup, and waste disposal can help motivate improved practices.

### Train Staff, Faculty, and Students

Training of researchers, students, and technicians who use hazardous materials helps protect human health and the environment. For most universities, employee training is required by OSHA's Hazard Communication Standards. Training must include general safety issues and more specific information tailored to the types of materials and protocols being used in order to reduce the frequency and seriousness of accidents and ensure that accident response is appropriate and timely.[3]

Successful training for environmental protection and safety associated with the use of hazardous materials can take many forms, although it should always be ongoing and interactive. Formal training in large groups can be cost-effective, but the content can be repetitious, resulting in high absentee rates. Training in smaller groups that focus on specific issues of a particular lab is more beneficial. These programs should be tailored to the needs of each laboratory or department to address specific risks, equipment needs, or procedures. Refresher courses for senior staff and researchers serve to update them on the latest information without requiring them to sit through lectures on information they already know. Hazardous materials management and safety training should be included in laboratory courses and required for students to work in laboratories, whether in a course, for independent research, or as paid assistants to faculty members.

Perhaps the most important and effective safety training occurs informally by the examples set by the faculty and research assistants. Some faculty and researchers ignore safety and waste handling rules because they feel very at home in the lab. For this reason, special attention should be given to faculty and staff who have the most contact with students so they consistently demonstrate proper ways of handling and disposing of hazardous materials and use of protective clothing and goggles.

Developing training can be expensive and time-consuming, but the benefits of avoiding accidents and violations of laws are self-evident. Training for environmental protection and safety needs to include the following:

- Personal responsibility and expectations
- Personal liability
- Location and purpose of Materials Safety Data Sheets
- Accident prevention
- Standard operating procedures
- Emergency procedures
- Disaster plans
- Lab inspections or audits (by EHS staff and self-administered)
- Review of federal, state, and local regulations

**Develop an Entrance and Exit Policy for Chemicals**

When new researchers join the faculty, they may bring research chemicals with them. Universities can benefit from policies to prevent problems resulting from the importing of chemicals in order to ensure proper inventorying, labeling, and dating of chemicals and supplies, as well as properly equipped laboratories and studios for the chemical hazard that is contained within. An entrance policy might limit and restrict the types and quantities of materials that are introduced by new faculty and require initial inspections and inventorying of the chemicals by the EHS staff.

When researchers leave an institution, they often leave their chemicals behind, and they may be the only ones who know the contents of unmarked, poorly marked, or old bottles. Cleaning out vacated laboratories can be very expensive, especially when the chemicals are unknown. In one university a retiring chemistry professor donated $20,000 to the university for new programs and equipment. Unfortunately, it cost the university $60,000 to clean out his laboratory because of unmarked chemicals. An exit policy strives to discourage faculty from leaving labs and studios with unlabeled chemicals and perhaps withholds the last paycheck until a final inspection is accomplished. Department chairpeople can be responsible for notifying EHS staff when a faculty member is leaving to ensure that final lab inspections are timely.

A similar exit policy can be established for graduate students who are conducting laboratory research. For example, the policy can require that students have their labs inspected prior to graduation and that the faculty member they worked for is held accountable for problems in the graduate students' laboratories, even if the work is largely unsupervised.

### Control Chemical Gifts

It is common for teaching and research institutions to receive free chemical products to evaluate. These samples may require special handling or costly disposal. The university can consider a policy that requires faculty to refuse samples or requires suppliers to take back unused portions. At the very least, the EHS staff should be notified about donated chemicals or other materials so they can be audited and tracked.

## Course and Research Design

According to RCRA, reducing wastes and hazards at their source is the preferred, although often overlooked, method of improving the safety of workers and the protection of the environment. Source reduction involves reducing the quantities or toxicity of hazardous materials by substituting safer materials or redesigning protocols. Source reduction initiatives can reduce risk to human health and the environment and reduce the cost of chemical inputs and waste disposal. Because they are hazardous waste generators, universities are required to have waste reduction or source reduction plans. In industry, source reduction initiatives in product and process design have dramatically reduced waste and improved safety.

Teaching techniques can dramatically reduce chemical purchases and resulting waste. Unfortunately, chemicals in academic research are often used for original research or research that must follow acceptable standard methods to be comparable to existing studies, regardless of the alternatives, so the potential for dramatic reduction from research redesign may not be extensive.

**Design Teaching to Use as Few Chemicals as Possible**

Microscale chemistry is a method of teaching the principles of inorganic and organic chemistry at high schools and colleges by using reduced amounts of chemicals. The teaching techniques were developed at Bowdoin College and Merrimack College in response to problems with air quality in the laboratory. The National Microscale Chemistry Center estimates that the conversion to microscale in colleges alone will eliminate nearly 4,000 metric tons of toxic waste annually and save hundreds of millions of dollars from avoided waste disposal alone.[4] In addition, microscale can decrease liability and accidents. It is now in use at over 2,000 colleges and universities worldwide and is gaining in popularity.

Microscale experiments use special small-scale equipment, so the major barrier to switching this form of experimentation is the capital cost of purchasing new equipment. A standard laboratory kit for a student in a beginning chemistry class can cost about $150. However, the decreased costs for materials and disposal of waste materials can usually pay for the new equipment in just over a year.

Increasing the use of instrumentation may also help reduce chemical use, waste production, or safety concerns in many case. New equipment may also reduce risk. For example, mercury thermometers and barometers can be replaced by equipment that contains hydrocarbons or is electronic, thus reducing the health risk from mercury exposure costs if a thermometer breaks.

Where it will not interfere with the quality of teaching, video presentations or demonstrations can be substituted for student experiments, reducing the amount of chemicals used in teaching laboratories. Using preweighed or premeasured reagent packets for student experiments helps to cut down on chemical waste, although it may increase solid waste and mixed waste (mixtures of solid and hazardous wastes).

**Design and Conduct Research Experiments to Reduce Waste**

Where possible, research projects may be designed and conducted to reduce hazardous waste. The same is true for developing photographs and painting. Some of these waste reduction strategies require substituting less toxic materials or eliminating materials altogether (e.g., the use of biodegradable scintillation fluids can replace those that are solvent based, and the use of predissolved acrylamide gel solutions can substitute for those mixed in the laboratory). Classes and research should try to reduce or eliminate the use of highly toxic or hazardous materials such as benzene, carbon tetrachloride, acetonitrile, mercury, lead, and formaldehyde.[5] In other cases, improved handling, careful measuring, and other commonsense techniques can reduce waste. The University of Indiana in Bloomington places the responsibility for waste collection on individual departments and stresses waste reduction in addition to waste handling. Each department must complete an annual report that outlines steps taken to reduce waste. The EHS department provides guidance on substitution of less hazardous chemical, inventory control, chemical exchange, and source reduction in research.[6]

**Incorporate Safety into All Teaching**

Universities need to incorporate applied safety rules and principles into the classroom, laboratory, and art studio. Incorporating these principles requires leading by example; teaching students about safety rules; holding students, teaching assistants, and faculty accountable; and maintaining safety equipment and supplies. Students may be required to take safety quizzes or answer safety questions as part of an exam. Safe behavior in the laboratory might also be considered in grading. The Joint Council for Health, Safety and Environmental Education of Professionals has prepared an excellent manual for helping instructors to incorporate safety into laboratory operating procedures and curricula.[7]

### Review Grant Proposals for Waste Reduction

University grant administrators or development staff routinely review applications for outside funding to ensure that things such as personnel benefits and university overhead rates are properly budgeted. This review might be expanded to include assurances that the funding proposals encompass the following areas:

- Identify the hazardous materials that will be used and the hazardous waste that will be generated.
- Describe what will be done to minimize the use of hazardous materials and subsequent generation of hazardous wastes.
- Include the cost of disposal of unused materials and waste as part of the budget.
- Specify what steps will be taken to reduce liability to the university and the risk of injury and environmental damage.
- Prevent the generation of mixed wastes (e.g., radioactive and solid waste or hazardous and solid wastes) which are difficult and expensive to dispose of.

The responsibility for reviewing proposals that include hazardous materials use may be given to an existing review committee or a new committee. This type of review is already common at most institutions where radioactive materials are used and experiments are done on human and animal subjects.

## Laboratory Facilities, Procedures, and Equipment

Good laboratory and studio facilities can help improve faculty and student research productivity, attract top-quality students and new faculty members, and reduce long-term liability. In addition, having sound operations and maintenance procedures can improve the safety in these facilities and the morale of the faculty, research technicians, and students.

### Ensure That Research Facilities can Accommodate Necessary Hazard Levels

University research, laboratory, and studio facilities should be equipped and designed to avoid accidents, handle accidents that do occur, have sufficient space to store all chemicals and wastes, and provide enough space for nonresearch activities, such as eating and meeting. These facilities are safest when they have the following features:

- Preventative equipment, including properly vented hoods,[8] backflow preventers, vacuum breakers, adequate storage space, and fire-safe storage cabinets for explosive and flammable materials
- Emergency equipment, including fire extinguishers, chemical spill kits, eye washes, and showers
- Properly functioning building systems, such as heating and air-conditioning that maintain a comfortable work environment. For example, when laboratories or studios are too hot, those working there may be inclined to wear inappropriate clothing (e.g., sandals) or fail to wear protective clothing, such as lab coats, gloves, or goggles.
- Conference rooms, student lounges, offices, and cafeterias to ensure that laboratory space is not used as a general work space or for eating

These facilities should be properly maintained to ensure the continued protection of human health and the environment, as well as to maximize research productivity. Safety systems and devices (e.g., eye washes, chemical hood ventilation systems) should be regularly checked and maintained. In addition, university administrators and researchers should clarify appropriate ways to pay for this equipment. Failure to find funds for these overhead items (either from administrative overhead or from research expenses) is a major factor in the failure to meet these essential needs.

**Require Protective Equipment**

The use of protective equipment, including goggles, gloves, and lab coats, is essential for personal protection and underscoring that the chemicals are hazardous. Sandals and shorts are not acceptable in the laboratory or studio, and long hair should be tied back. It is especially important that everyone, even faculty, working in or passing through a laboratory or studio wear the appropriate protective clothing and equipment. This expectation is aggressively enforced in industrial settings, and similar enforcement should be undertaken at universities. As in industry, laboratories should have adequate goggles and other necessary equipment for visitors and forgetful technicians and students, so there is never an excuse for noncompliance.

Any time that a photograph in a university publication shows students, staff, or faculty working in a laboratory, they should be wearing protective apparel.

### Conduct Regular Inspections

Regular (semiannually) audits of laboratories and studios using hazardous materials should be conducted in order to identify problems, locate areas for improvement, and create baselines against which future waste generation can be measured. These inspections are usually carried out by EHS staff or representatives from the safety committee or by colleagues. Inspections should identify source reduction opportunities, as well as hazardous conditions. Faculty, researchers, and students need to understand the purpose of audits and be informed of the results and necessary follow-up steps. Between formal audits, research and teaching staff should conduct comprehensive self-inspections.

### Require Safe Conduct

Safety rules should be followed at all times and by all who use labs and studios where hazardous materials are used. The following rules are important:

- No eating or drinking in the laboratory or studio is allowed.
- Never work alone.
- Never burn plastics. The phosgene gas resulting from burning plastics with a chlorine polymer can cause breathing difficulty.
- Always store highly toxic or dangerous hazardous materials in nonbreakable secondary containers.
- Always transport hazardous materials in a nonbreakable secondary container.
- Always purchase the smallest-volume containers that are reasonably possible.
- Report all spills and accidents.[9]

### Make Materials Safety Data Sheets Readily Available

Materials Safety Data Sheets (MSDS) provide basic information about a chemical or product and include hazards and precautions for handling. Workers have a right to be fully informed of the chemicals used in the workplace. To accomplish this and make safety information available, MSDSs should be available to everyone who works with chemicals. This can be done with on-line computer databases or a notebook in the lab or studio. Training programs should include the interpretation and use of MSDSs.

### Design Cleanup Procedures Carefully

Using appropriate cleanup practices can reduce waste. For example, nontoxic detergents are preferable to chromic acid or acetone for cleaning glassware, and some soaps and bleach solutions used for cleaning may contain high levels of mercury. In addition, the segregation of wastes, with appropriate and well-labeled containers, and prompt collection of those wastes from the lab can reduce accidents and improper disposal.

Careful measurement and proper maintenance of measuring apparatus can ensure that proper quantities are used and waste does not result.

### Restrict the Transportation of Chemicals and Hazardous Wastes

In order to prevent spills and comply with applicable regulations, university departments should strictly observe safety precautions in the transport of chemicals. Inside a building, chemicals should be transported inside a secondary container that can catch spills or contain a spill in case the material is dropped or hit. Transporting chemicals between buildings requires compliance with U.S. Department of Transportation regulations and special packing. Students should not be allowed to transport chemicals between buildings unless they are appropriately trained, overseen, and insured.

The rules governing the transport of chemicals and wastes depend on the amount of waste that the institution generates. Because only institutions generating very small quantities of waste are exempt from these regulations and transporting chemicals can result in broken containers or spills, always follow all applicable regulations when transporting chemicals.[10]

## Chemical Purchasing and Storage

Appropriate storage and inventory control can help avoid overordering, purge outdated material, and ensure safety. Chemical storage and management are most often done on a lab-by-lab basis, but cooperating with the purchasing department by centralizing the purchasing of chemicals, particularly chemicals used in very large quantities or by many labs, may offer increased efficiency and safety and reduced costs. Well-designed purchasing systems can help promote proper chemical inventory management, ensure that a laboratory is equipped for the hazards within it, determine whether the desired chemical is available from within the university, and improve source reduction. Nevertheless, relying on the purchasing department to encourage or enforce safety and source reduction policies can be problematic since researchers may view the department as a bureaucracy that slows them in obtaining needed supplies and does not understand how and why chemicals are used. The university environmental committee or the safety committee can be instrumental in working through these conflicts. (See also chapter 4.)

**Undertake Responsible Purchasing**

The purchase of chemicals and the negotiation of contracts with chemical suppliers can reduce hazards and unnecessary purchases. However, many of the purchasing decisions for hazardous materials are made at the department level by faculty, teaching assistants, or stockroom managers. The American Chemical Society estimates that as much as 40 percent of the hazardous waste generated in science labs results from excess chemicals or chemicals that have expired.[11] Bulk purchasing, which is usually good for the environment in order to reduce solid waste, is generally ill advised when purchasing chemicals, since disposing of excess chemicals is costly. Purchasing practices should account for the cost of waste disposal as well as the purchase price.

**Store Chemicals Properly**

Proper chemical storage can prevent accidents, since some chemicals are incompatible and form poisonous gas, corrosive solutions, or fire hazards if combined. In addition, orderly and logical storage will facilitate inventory management. Some university laboratories store chemicals in unsafe containers, on corroding shelving, or with incompatible chemicals stored together.

Safe chemical storage requires knowing which chemicals are compatible and, especially, incompatible; providing proper facilities, including storage cabinets for flammables, noncorrosive cabinets for acids, and explosion-proof refrigeration for certain chemicals such as ether and hydrogen peroxide; and understanding the hazards of particular chemicals such as nitric acid and ether. Each laboratory must designate an individual to ensure that storage is proper and that faculty and students using the storeroom understand how to find and store chemicals. Information about proper chemical storage is available from the EHS staff, chemical suppliers, or state and federal environmental protection agencies.

### Label Chemicals Properly

Stored chemicals must be accurately labeled with the chemical name and formula, the date of expiration, the date a reagent was created, and any special considerations. Information about chemical characteristics, classification, compatibility, and special storage should also be included on the label. Labels from chemical suppliers should be left on the bottles. Labels should be affixed with tags that resist chemical drips and will not fall off. University laboratories are full of experiments, and it is possible to find bottles that are labeled "mixture A" or "Steve's reagent." Disposing of unknown chemicals such as these is expensive since the material must be assumed to be dangerous. Safe storage and transportation of a chemical also requires knowledge of its properties and hazards.

### Change Purchasing Forms to Reflect Source Reduction Policy

The order forms used to order chemicals can serve as a reminder of the institution's source reduction policy and even as a mechanism for enforcing the policy. The chemical purchasing forms can include a reminder about purchasing only as much as is needed at the time (for example, no more than a three-month supply) or prohibited practices (e.g., flammable liquids are prohibited in containers larger than one gallon). For example, the chemical purchasing form might include questions that the faculty member or stockroom technician must answer:

- Will this chemical be used up in less than three months? (Requires special consideration if longer than three months.)
- How will this chemical be disposed of?
- How will this chemical be stored?

### Develop an Inventory System

An accurate inventory of the chemicals and wastes in labs, departments, and stockrooms is required by OSHA for safe and responsible management. An inventory can prevent overordering, identify expired chemicals, and help ensure that proper storage procedures are followed. At the College of DuPage in Illinois, a chemical reduction program of inventory control and source reduction in the laboratories reduced chemical orders by forty percent.[12]

The burden of inventorying tasks can be eased by working with the purchasing department to collect information about the materials as they are ordered. Although most universities inventory chemicals by hand, technologies such as computerized bar code systems allow easy recording of information on chemical names, quantities, location of initial and final storage, when and from whom the chemicals were purchased, who the actual user was, and how the residuals were disposed. The system can generate regular reports for departments (e.g., worker right-to-know information, waste minimization reports) and identify any potential or actual problems, such as expired chemicals or chemicals that are more hazardous than the facilities can accommodate.

With the bar code system, inventory information is gathered using a laser scanner and a portable data pack. The operator notes his or her location by entering a numerical identifier into the data pack and then uses the laser scanner to read the bar code labels on each container. The amount of each chemical is determined by eye, and a code (for full, half full, etc.) is entered from a menu. Once the information is gathered, it can be downloaded into a database.

A bar coding system requires an initial investment of equipment (although a university may have bar code inventory systems for tracking other equipment) and time to computerize the initial inventory information. Most chemical suppliers provide bar-coded labels on their chemicals and can electronically provide the database of information needed to get started. Box 7.2 shows the types of information that can be included in a bar coding system. Once a bar coding system is up and running, it can save much of the time needed to conduct inventories by hand.

### Centralize Chemical Ordering and Storage Where Possible

A well-stocked and well-run central stockroom for frequently used materials may reduce laboratory inventories, improve inventory and storage practices, reduce chemical spoilage, and facilitate chemical sharing. The central location may serve an entire department, such as chemistry, or

**Box 7.2**
Information typically contained on a chemical bar code

1. Bar code number.
2. Type of agent. Is this agent a trade name substance or a pure chemical? Is it a physical hazard or a biological hazard?
3. Hazardous agent. The chemical, physical, or biological hazard that is being added to the inventory.
4. Supplier. The manufacturer or supplier.
5. Catalog number. The supplier's catalog number.
6. Quantity and unit of issue. The quantity and size of the chemicals originally ordered.
7. Amount on hand. The amount of the original order still on hand.
8. CAS number. The Chemical Abstract Services' unique assigned number for this substance.
9. MSDS. Is there a Materials Safety Data Sheet on hand for this chemical (yes or no?)?
10. Grade. The particular grade of the chemical (for use with pure chemical agents).
11. Acquisition date. The date the chemical was acquired.
12. Expiration date. The date on which the chemical use will expire.
13. Activity on acquisition. The activity of radioactive substances on the date of acquisition.

the entire campus. When run by skilled staff, centralized storage is easier to manage and inventory than individual laboratory or departmental storerooms. Although individual labs are likely to retain their own chemicals, centralization can reduce the number and quantities of chemicals stored in these research labs. A central stockroom can also promote pollution prevention techniques. For example, they can distribute nonhazardous cleaning solutions rather than acetone or chromic acid, which must be disposed of as hazardous waste. To be effective the storeroom needs to be located close to the labs it serves or be equipped to transport hazardous materials safely, legally, and quickly.

### Establish a System to Exchange Chemicals

Maintaining a list of which excess chemicals are available can provide for exchanges between researchers or reuse of material instead of disposal. If chemicals are exchanged between buildings, federal regulations for packaging, labeling, and transporting chemicals must be followed. Some university systems operate exchange newsletters to assist the exchange of unused or unwanted hazardous materials within departments. In a single stockroom, this can be safely accomplished by affixing a designated and dated sticker to unwanted hazardous materials that signifies that it is "up for grabs" by authorized faculty and students.

## Waste Disposal

Although reducing the wastes generated through source reduction measures is the most effective way to reduce disposal costs and risk, hazardous wastes are still generated in laboratories and some studios. Many of these waste products are regulated and must be disposed of as chemical, biological, radioactive, physical (including glass and sharp objects), or mixed wastes. Failure to comply with regulations can result in fines and future liability for both the researcher and the university. Most important, proper disposal of hazardous waste generated from research ensures the protection of human health and the environment outside the lab.

Educational institutions generate small quantities of many different types of wastes (in contrast to industry, which tends to generate large quantities of a few materials). Waste generation in universities is sporadic and influenced by the types of activities, the beginning or ending of projects, the end of a semester or term, donations of unneeded material, the accumulation of wastes from several years or users of a lab, and relocations of researchers from one lab to another.[13] University wastes are typically packed together in "lab packs," fifty-five-gallon drums that contain up to fifteen gallons of liquid wastes in bottles or cans and separated by absorbent filler that reduces the possibility and hazard of breakage. The lab packs are transported to a disposal facility, where the wastes are segregated and treated or landfilled.

### Know What Are Hazardous and Regulated Wastes

Hazardous wastes are defined by RCRA as materials that are corrosive, such as acids and bases; toxic, such as mercury and lead; ignitable, such as oil and organic solvents; reactive which are unstable and can cause explosions; and wastes that are specifically listed in regulation. Radioac-

**Table 7.1**
Hazardous wastes generated in college courses

| Course category | Potentially hazardous waste |
|---|---|
| Agricultural sciences | Pesticides, fertilizers, pharmaceuticals |
| Allied health services | Chemicals, pharmaceuticals |
| Biology | Chemicals, lab supplies, protective clothing |
| Chemistry | Chemicals, lab supplies, protective clothing |
| Communications | Photochemicals, lubricants, oil |
| Crafts and design | Paints, solvents, inks |
| Engineering: civil, mechanical, electrical | Chemicals, waste oil, solvents, degreasers, PCB-contaminated electronics |
| Environmental controls | Charcoal filters |
| Fine arts | Paints, solvents, metals, inks |
| Industrial production | Solvents, degreasers, oil, grease, acids, bases |
| Life sciences | Chemicals |
| Medical | Chemicals, protective clothing, biomedical, biohazards |
| Psychology | Chemicals, solvents, biomedical |
| Physics | Chemicals |
| Visual and performing arts | Paints, solvents, developing and fixing chemicals |

Source: U.S. Environmental Protection Agency *Report to Congress: Management of Hazardous Waste from Educational Institutions* (U.S. Government Printing Office, Washington, DC: 1989).

tive wastes are not defined as hazardous wastes but are regulated nonetheless. Regulated wastes are generated throughout the university, with the highest concentrations in research and teaching laboratories.[14] Wastes are also generated in psychology and engineering departments, art studios, photographic developing labs, electrical shops, and the buildings and grounds department. Universities with hospital or biomedical facilities generate biohazard wastes. Table 7.1 shows the wastes that are likely to result from college-level courses. Hazardous waste also includes contaminated protective clothing used in handling chemicals. Radioactive material is handled separately.

Federal and state law regulate generators of hazardous wastes differently depending on the amount of wastes that they generate. Those that generate below a certain total quantity of waste may be exempt from some regulations and reporting requirements. The fact that the institution may become exempt from some regulations if it reduces its generation of wastes should be a powerful motivator for waste reduction. Complete and up-to-date discussion of the applicable regulations is available from the EPA or state department of environmental protection.

**Designate Responsibility for Hazardous Waste Disposal and Tracking**

Certainly safe practices and waste reduction must be everyone's responsibility, but the management of hazardous wastes must be the designated responsibility of an individual or department in order to ensure that the university is in compliance with regulations. Waste disposal requires selecting and monitoring contractors, securing needed permits, and tracking a seemingly endless stream of paper since each waste has a multicopy manifest that documents its path to its ultimate disposal. Failure to track the manifest or inspect and record other required information can result in large fines even if laboratory conditions are exemplary.

**Dispose of Hazardous Wastes Appropriately**

Proper disposal of laboratory wastes, and the associated storage, handling, and tracking of these wastes, is complicated. Detailed regulations are provided by state departments of environmental protection.

Regularly scheduled waste pickups and waste disposal efforts can help reduce stockpiling of unneeded chemicals and ensure that expired chemicals are disposed of properly. In contrast, when waste disposal is conducted on an as-needed basis, there is less incentive for keeping updated inventories or managing wastes on an ongoing basis.

Some materials can safely be poured down the drain; this disposal method, however, is used far more often than is legal or safe and may disrupt the sewage treatment or seriously contaminate waste water. Wastes that can be poured down the drain must meet requirements for pH levels and concentration of metals. Other means of disposal are by incineration, treatment, or storage in a hazardous waste landfill.

Many university researchers pretreat hazardous wastes by neutralizing strong acids or bases so they can be poured down the drain. In most states, however, these practices fall into a regulatory gray area or are prohibited.[15] The practice of pretreating wastes in the laboratory should be closely monitored or avoided altogether.

Since there are few facilities for the disposal of radioactive wastes, universities may be forced to decay low-level radioactive waste on site. Proper facilities and training must be found to accommodate this need. Other lab wastes, such as animal carcasses, should be handled and disposed of in compliance with the relevant regulations.

**Select a Waste Disposal Contractor Carefully**

As a generator of hazardous waste, the university is responsible for ensuring that its wastes are disposed of legally, even after the wastes have been handed off to a disposal contractor. Universities (usually the EHS person) should check with their state department of environmental protection to make sure that the contractor who transports and disposes of the university's hazardous wastes has the necessary permits. If the contractor subcontracts with a second company, the university must still ensure that the second company is permitted and in compliance with the permit conditions and regulations. The background check of contractors should determine if they are financially stable and have the appropriate pollution and liability insurance. Whenever possible, a university representative can regularly conduct site visits of a disposal contractor's facilities.

**Determine How to Pay for Waste Disposal and Safety Equipment**

The university must ensure that disposal costs are covered in full, by either individual academic or research budgets or an overhead budget (e.g., the School of Arts and Sciences pays for the disposal of wastes generated in any of its departments). Failure to fund disposal fully can result in improper storage and disposal and associated risk and liability.

Charging the generating laboratory or studio or its department for the actual disposal can provide an incentive for waste reduction by providing a connection between waste generation and the cost of hazardous waste disposal, but this strategy may cause staff to cut corners and dispose of wastes improperly in order to realize the savings. Charging waste disposal costs back to the generating laboratory, studio, or department can also create incentives for improper or lengthy storage of material that increases the risk of fire, associated fines, or violations of wastewater permits as material is improperly disposed of (e.g., down the drain). A central budget for waste disposal will help to overcome these incentives for improper waste handling and disposal. Another possibility for reducing purchase of unnecessary chemicals is to add a disposal tax to the price of the chemicals when purchased. Notre Dame University is trying this approach.[16]

Grants are major sources of funding for many laboratory research projects in science and engineering. However, the process by which grants are awarded frequently ignores the consequences of the research on safety, health, and the environment. Each granting agency or organization has a unique application and review process, and often grants provide funds for supplies such as chemicals, but many do not allow the project to use funds for waste disposal and safety equipment. (U.S. Army grants are one exception.)

Federal grants to universities usually have a negotiated overhead rate that includes payment for laboratory space, electricity, water, heat, and university services, such as payroll and human resources. In some cases, the overhead charges specifically include waste disposal and safety equipment costs; in other cases these items are excluded or payment for these necessary expenditures is uncertain. A university's internal grant review process, ordinarily responsible for ensuring that financial and contractual information is accurate, must see to it that environmental and safety-related costs are covered by the research proposal itself, the overhead that the university negotiates with the granting agency or foundation, or some other means. The grant review process should also ensure that the facilities where the research takes place can adequately handle and store hazardous materials. This grant review can be done at the university level by the grant administration staff or at the department level. The EHS committee should help determine the appropriate method for the university.

Some researchers are reluctant to add the cost of waste disposal and safety equipment to their research budgets, fearing that these costs will be considered overhead items by agencies or foundations evaluating the research and will put the proposal at a competitive disadvantage. Most institutions do in fact consider these disposal costs as overhead, although some universities (e.g., the University of Rhode Island) pass waste disposal costs back to each research project. It is important for universities to recognize that safety equipment and waste disposal are essential costs of laboratory research and ensure that the issue is addressed before improper handling creates problems.

## Art and Photography Studios

Awareness of chemical hazards is relatively high in science and engineering laboratories but often quite low in art studios and photographic developing laboratories. Similar safety and environmental precautions, especially for the disposal of wastes and inhalation of fumes and dust, must be observed. Preventing exposure, reducing the generation of waste, and ensuring the proper disposal of hazardous wastes are essential.

### Recognize That Hazards Exist

Art studios and photography labs may contain solvents, aerosol sprays, gases (e.g., oxygen, acetylene, and propane), dusts, dyes and organic pigments, metals and metal dusts, and strong acids and bases, all of which can cause harm to people and the environment if inhaled, touched, or spilled. Art departments can address these issues with the help of EHS staff and as participants in the campus-wide environmental committees.

### Take Steps to Reduce the Hazards

Paints, paint thinner, stains, dyes, and enamels may contain methanol and other solvents that are rapidly absorbed by inhalation and skin exposure. These substances should be used only in well-ventilated areas. Because they are hazardous materials, alternatives should be sought. Art departments can reduce the use of oil-based paints by substituting latex paints and undertaking aggressive solvent settling and reuse. Disposal of waste paint and paint thinner must follow procedures for the disposal of other hazardous wastes. The EHS staff can assist art departments and should include them in all auditing initiatives, waste disposal programs, and EHS committees.

Silica flour is commonly used in art classes, and silica dust may be produced when grinding and polishing pottery. The dust can cause respiratory inflammation and scarring of the lungs. The resulting disease, silicosis, is progressive and can result in continued scarring and shortness of breath even after exposure has ceased. Containing the silica dust through proper ventilation and the use of fume hoods can reduce exposure.[17]

Art students may also be exposed to asbestos dust from frayed gloves and aprons or insulation in and around kilns and ovens. French chalk dust, used to dust lithographic plates, and serpentine stones such as soapstone and African wonderstone also contain asbestos. Whenever possible, exposure to asbestos should be eliminated since asbestos dust penetrates and damages the lungs. Respirators should be worn when exposure cannot be avoided.

**Recycle Photographic Chemicals**

The disposal of chemicals from darkrooms poses a hazard to the environment. The most problematic element is silver, found in spent photographic fixer in concentrations of several thousand parts per million. Silver is a heavy metal that interferes with the ability of bacteria to break down biological material in sewage treatment systems. The Massachusetts Water Resources Authority, for example, has an established limit of 2 parts per million (ppm) of silver for effluent entering the public sewage treatment system. Under RCRA, solutions with more than 5 ppm of silver are regulated. Silver is also a precious metal that can easily be recovered and sold on the market.

There are several options for reclaiming fixer collected in the darkroom. The first option is to hire a licensed disposal company to collect and remove the fixer solution regularly. If the university is a large-quantity generator of hazardous waste,[18] the fixer cannot be stored on campus longer than ninety days, even if the fixer quantities are small. A second option is on-site treatment by using a reclamation unit—an electrolytic machine that extracts the silver from the waste fixer. The resulting effluent can be discharged down the drain and the silver sold. If the university uses large quantities of fixer (over about 300-400 gallons per year), it is probably cost-effective to purchase a reclamation unit, although the burden of permitting and overseeing the unit can be time-consuming. Wastes from darkrooms across campus can be consolidated if they are transported legally.

## Conclusions

A variety of hazardous materials are used and hazardous wastes generated in college and university research and teaching laboratories, art studios, and photography labs. Many of these materials are integral to the business of education and research; nevertheless, the risks associated with their use must be reduced. Many of the desired strategies will require rethinking experiments, research, and the supporting laboratory stockroom procedures. The benefits accrue in reduced accidents, liability, and regulatory burden.

University teaching and research must be conducted in a way that protects human health and safety and the environment. The principle of preventing pollution before it occurs can successfully be applied to academic research design, execution, and cleanup and is the best strategy for reducing safety risks, liabilities, and associated costs. In labs and studios, pollution prevention means paying attention to reducing the quantities or toxicity of hazardous materials before they are generated, using appropriate facilities, developing policies and training, and using waste disposal systems.

The strategies described in this chapter are some of the essential pieces of a successful overall pollution prevention approach, but are incomplete without a commitment to the very principles of reducing risk to health, safety, and the environment.

# 8

# Student Activities

As the primary customers of educational institutions, students are in unique positions to influence their universities' environmental stewardship activities. Students and their activities result in the purchase of goods, the generation of wastes, and the use of resources, including water and energy. Since many students live on campus, their living habits and choices influence the college's own environmental footprint. Studies of waste from student residences show that its makeup is similar to that of the university as a whole (about 50 percent paper) but that it varies a great deal depending on the day of the week and with the time of year. Student preferences influence food choices; colleges where more students live and eat off campus tend to have more take-out eating establishments and thus more solid waste from dining services. Lights, heat, and water are used lavishly in nearly every college dormitory.

Many students discover a new sense of independence when they enter college. Often it is the first time that they can make their own choices about issues such as how they spend their time, what they eat, how they get around, and how long to shower. Despite student interest in and concern for global environmental issues, it is often difficult for them to see their personal connection to environmental issues and to become motivated to take action. Most undergraduate students are only beginning to understand the real world of cost trade-offs, compromises, and service.

University student population turnover—about a quarter of the undergraduate students graduate each year—brings new energy and ideas but often drains administration and staff energies as the same environmental issues are raised year after year. For example, each year managers of Tufts dining services must explain that student theft forces them to use disposable dishware in the campus take-out cafeteria. Administration and staff wrestling with meeting budgets, ordering chemicals, and getting thousands of meals served can find student naiveté either irritating or energizing. The green university can find ways to benefit from the influx of new students' energy to research or implement change, communicate past findings and progress, and encourage new ways of attacking old problems.

## Advocates for Environmental Change

Students are working for environmental change on many campuses and in many communities. They have created campus recycling programs, organized reusable mug programs, collected used books and clothing, distributed compact fluorescent light bulbs, picked up litter, planted trees, and much more. Student leadership and advocacy can address single issues, such as the effort by Tufts graduate students to lobby the dean of the graduate school to allow master's and Ph.D. theses to be printed double-sided and on recycled paper in order to reduce waste. On an ongoing basis, students can also be effective advocates promoting environmental initiatives such as source reduction and recycling programs by conducting routine and regular training and updating department secretaries, dorm monitors, and other staff about these programs. Some student initiatives are begun independently, and others are coordinated with other university efforts or with efforts at other schools. For example, National Wildlife Federation's Campus Ecology Program's four field coordinators and senior staff can help interested students and staff find resources, implement change, and promote their programs. The Campus Ecology Program provides technical assistance, coordinates regional and national training clinics, and produces special reports and educational materials on target topics. The program has provided assistance to student-initiated environmental audit projects at SUNY-Buffalo, New York, Bemidji State University, Minnesota, Warren Wilson College, North Carolina, Princeton University, New Jersey, and at numerous other campuses throughout the country.

There are examples of student leadership beyond their own campuses as well. Students at Yale University raised funds for and staged the Campus Earth Summit in 1994. This summit, which convened university students, faculty, and staff members from campuses nationwide, examined the need for incorporating environmental knowledge into all university curriculum and environmental stewardship action into campus operations. Following the summit, the students distributed *Blueprint for a Green Campus.*[1] The Student Environmental Action Coalition organized its national conference for environmental stewardship several years ago, providing many campus groups with the energy and empowerment to undertake environmental stewardship efforts (often in the form of environmental audits) on their own campuses.[2]

**Serve as the Environmental Conscience**

On most campuses students feel freer than faculty and staff to criticize administrative decisions and actions. This freedom allows them to serve as a university's environmental conscience. Students who take on the role of an institution's environmental watchdog often discover that they must take care to build rather than burn bridges with the faculty, staff, and administrators.

Students participating in the nationwide student movement to conduct environmental audits of campuses have benefited from working directly with administrators and staff to solve problems rather than exposing problems in the press or with inappropriate or embarrassing publicity. On the other hand, well-placed publicity can help draw attention to a problem, especially if the problem is ongoing or ignored. In 1993, students at a number of schools in the Northeast, including Dartmouth, Tufts, and Wellesley, successfully lobbied their boards of trustees to disinvest university holdings from Hydro-Quebec's James Bay projects, which would have disrupted the fragile ecosystems of northern Canada and displaced native peoples. The students in this campaign served as an environmental consciousness and called on the boards of trustees in the schools to recognize the consequences of their actions. The student initiatives were effectively coordinated among campuses with persistent and professional protests, letter-writing campaigns, seminars, and meetings with college presidents. These efforts began slowly and grew to a regional campaign that was well organized on and off the involved campuses.

### Influence Student Government to Promote Environmentally Friendly Choices

Student governments have a great deal of autonomy on most campuses. Furthermore, student governments often distribute funds to student organizations and sanction events throughout the college or university. Consequently, there are numerous opportunities for them to leverage their funds in ways that are beneficial to the environment. For example, student governments can pass resolutions requiring the student organizations they sponsor to do the following:

- Use recycled paper.
- Fund and promote carpooling programs to special events or at holidays.
- Purchase energy-efficient computers and other equipment.
- Limit the use of disposable dishware at outdoor special events.
- Recycle material generated at student government–funded events.

In the same way that environmental stewardship efforts need to be linked with university priorities, these efforts will be most successful if they also build on existing student priorities. For example, if student government is encouraging community service, student environmental activists will be more effective if they focus their environmental message on service rather than responsibility.

### Take Personal Choices Seriously

Despite the fact that environmental issues are important and diverse, not all students are inspired and motivated to take action. Fellow students can be powerful motivators for change by providing information to their peers, serving as examples, discussing and debating issues, and demonstrating their importance. Students are often surprised to discover how influential their own actions are in motivating their fellow students, university faculty, and staff. Personal action to take public transit, turn off lights and computers, recycle, and reduce waste by carrying a reusable bag, cup, and flatware are meaningful examples that are noticed daily. On one campus, an environmental leader picks up two pieces of litter each time she walks between campus buildings; the effort elicits comments, praise, thought, and action by others. The strategies that are appropriate on each campus are different, but the results of student-led campaigns can be impressive.

Motivating students to take personal action requires linking environmental protection issues with the diverse issues that students care about. For example, child studies majors may be motivated when they understand how air pollution can affect children's health and ability to learn, and budding political scientists may become interested when issues of hazardous waste policy are linked with those of race and class.

Students employed by the university or college have unique opportunities to influence staff, try new initiatives, and demonstrate how environmental commitment and action can complement existing systems and objectives. For example, students who work in the library can improve the collection of white paper in the offices, and students working in the dining halls can shut off lights between meals. These jobs are particularly important examples, since motivated students can demonstrate how environmental stewardship need not detract from accomplishing job responsibilities. Students in these positions can be important trainers of their coworkers and can provide valuable feedback about the pragmatic aspects and needs of any program.

### Use Student Media to Promote Eefforts

Students have many opportunities to influence their peers, the administration, and the faculty through student papers, TV shows, and radio. Regular stories and public service announcements can be effective ways of legitimizing and raising awareness of the environmental efforts around campus and asking for ongoing participation. Regular columns in the student paper or spots on radio talk shows that focus on the environment can educate and motivate.

## Students as Participants in Change

Students are valuable in the effort to see that environmental projects are undertaken. In the busy world of college administration, building management, food services, and purchasing, there is little extra time, and environmental initiatives often take a back seat to the business of academic and operational departments. Students can be effective extra hands to see that the research, testing, and implementation of environmental action take place.

Efforts to improve environmental performance will be longer lasting if they are woven into the fabric of university life and complement university priorities. However, sometimes students are most effective when they are responding to a need that they identify themselves or when they take the initiative to develop and execute projects independent from university objectives. University environmental efforts are well served if they can capitalize on student enthusiasm and energy rather than trying to manufacture it. There are countless examples of student initiatives that were organized outside mainstream campus environmental efforts or where students have developed innovative solutions to campus problems. For example, Tufts' Environmental Consciousness Outreach (ECO) students organized the Free Sale, an event to encourage the reuse of unwanted clothing, games, books, and furniture. Students at the University of New England initiated a book collection and shipped cartons of books overseas. At Harvard, students initiated one of the first dormitory competitions to save electricity, conserve water, and improve recycling. Student environmental leaders organize these types of events, but they are successful because many students participate.

**Form Environmental Committees**

A campus-wide environmental committee can provide valuable links between student initiatives, energy, and advocacy, and the faculty and staff. A committee can provide students with a forum to raise environmental concerns. Likewise, a manager from an operations department can coordinate with students at committee meetings.

Students often have environmental committees or groups that are independent of the university-wide committee. These groups address environmental issues on and off campus, host environmental speakers, and conduct volunteer initiatives. Sometimes these groups are interested in working on campus-based environmental stewardship, yet they may have other agendas. University staff and faculty should keep in close touch with these groups.

**Use Student Volunteers or Employees**

University administrators and staff can create short- and long-term work-study student jobs to assist with environmental initiatives as researchers, extra hands and muscle, or participants in department decision making. These relationships may be quite different from simply employing students in dish rooms or mailrooms or to shovel sidewalks. Although some students lack the experience needed to implement projects on their own, they can assist departments in researching new technologies that improve efficiency or reduce waste, performing inventories and audits to inform environmental actions, spearheading educational campaigns, and testing new products. These work-study jobs cost-effectively add staff resources and provide learning and professional experience to students.

Students from environmental or service clubs are valued resources for staff, faculty, and other students who are carrying out these environmental projects. Finding student volunteer opportunities or student volunteers is not always easy. At Tufts, we found that short-term, well-defined activities, such as collecting recyclables after a special event or publicizing a contest, were easier to recruit for than long-term projects. More dedicated students were willing to serve in an ongoing capacity as members of a committee or regular water meter readers for our competition between dormitories. Graduate students were instrumental in conducting in-depth investigations and developing working pilot tests of proposed programs.

**Combine Curriculum with Hands-On Practice**

Campus environmental initiatives provide interesting and exciting educational material for learning. Students can examine resource use, new technology, purchasing decisions, and policies for their academic papers and research projects. Faculty can organize campus-based environmental topics to teach foreign languages, economics, engineering, and computer science. The key to developing successful topics is to narrow the scope of the projects sufficiently so that their results are informative and useful to university decision makers. Too often academic assignments examine issues with such breadth that students learn concepts but cannot apply them. For example, instead of evaluating energy-efficient technologies available to use on the campus, students and faculty can narrow the project scope to explore the details of how their university would switch from conventional water heaters to a solar hot water system in a specific building. Students will find that their recommendations are more likely to be followed if university staff are involved in the analysis and consulted prior to developing final results, much in the way a consultant or an architect works with a client throughout a project.

Student academic projects that are hands-on projects rather than written reports are also more likely to result in change. For example, electrical engineering students might perform a lighting audit and install new lighting controls on a pilot basis, or chemistry students might help the environmental health and safety department to evaluate unknown wastes or manage chemical wastes.

**Use Peer Outreach to Make Initiatives Succeed**

Many environmental initiatives depend on modifying the behavior of individuals by using education and training. Students can be effective trainers and are particularly effective in reaching their peers. In many instances, staff and faculty are more open to student trainers or students who carry a personal message of environmental commitment than they are to a similar message from staff. On some campuses, members of a student environmental group have fanned out across campus to introduce a new recycling program or describe a new bus service. This one-on-one education helps to build successful and long-lasting programs.

University managers who want to reach the diverse university community can benefit from partnering with students. For example, a purchasing manager who wants to make academic departments aware of new recycled products could provide samples to interested students, who could visit department managers to show and discuss the new material.

**Seek Student Input**

In the same way that students who strive to motivate environmental action on campus should seek input from university staff and administrators, staff and administrators should also seek input from students. Since students infiltrate all parts of the campus and use its services at all hours, they have relevant and insightful input on systems that affect them personally and the campus as a whole. For example, a student survey might

**Box 8.1**
Personal environmental action in university residences

Turn off lights and equipment when not in use.
Don't use the toilet as a trash can.
Take short showers.
Wash full loads of laundry.
Reuse scrap paper.
Recycled used cans, bottles, and paper.
Cook efficiently by covering pots and using the smallest burners.
Don't preheat ovens longer than needed.

find that the barrier to increased use of a campus shuttle is not the schedule but lack of information about the schedule itself.

## Reducing Impacts in Residence Halls

Dormitories and other student housing generate all of the environmental impacts found in offices and single-family homes: solid waste, water pollution, air pollution, and toxics use. As in other buildings, the condition and operation of the dormitory play a large part in determining the degree of these impacts. The residents' habits and practices too can change the environmental footprint.

### Residents Take Action

Residents of dormitories and other university housing facilities have opportunities to use their living facilities in ways that are wasteful or environmentally friendly. Actions are shown in box 8.1. Some students will be motivated to take additional action, perhaps in houses where they cook for themselves or in a campus "environmental house." These actions might include selecting organic foods in their own cooking, on-site composting with worm bins, or a housewide commitment to water conservation. Environmental actions by college dormitory residents can be discouraged when students feel that the high cost of room and board entitles them to all the electricity, heat, water, and waste generation they want. University housing coordinators and student environmentalists can be instrumental in educating fellow students about the financial and environmental costs of twenty-minute showers, overheated rooms, and excessive trash.

### Install Energy-Efficient and Centralized Equipment

In university residences, where laundry, stoves, ovens, and refrigerators receive heavy use, energy-efficient appliances can have rapid paybacks. For example, new super-efficient refrigerators, developed with financial support from the EPA, will be commercially available soon. In contrast, minirefrigerators, common in dormitory rooms, are extremely inefficient. University housing offices should discourage or even prohibit them and provide central refrigeration for student foods. Individual student bins or bags in the central refrigerator can help reduce the tendency to view all food stored there as communal. The university housing offices can decline offers from outside companies that rent inefficient refrigerators, microwave ovens, and other appliances to individual students because the increased electricity use makes them financially and environmentally costly. Halogen torchère lamps that use 300 watts or more should be banned for safety and energy reasons.

Water-conserving washing machines and dryers are now available at reasonable prices. Given the intensive use that laundry machines receive in college dormitories, they will pay for themselves quickly in terms of water, water heating costs, and electricity or natural gas. Changing to water-conserving laundry machines must be accompanied by student education about how to use the new machines since less soap is needed for each load. College laundry machines can also be adjusted to provide only a cold water rinse, thereby reducing hot water use.

### Provide Information and Feedback to Dormitory Residents

Research has shown that feedback and information about personal resource use can have a positive effect on behavior.[3] Several years ago, the University of New Hampshire's energy manager spearheaded a "flip the switch" campaign. His office printed out monthly graphs showing how the recent month's electricity use compared with previous months and with the same period in preceding years. The campaign reduced electricity use measurably and created a means by which people could see the measured results of their efforts.

### Residents as Building Stewards

A single dripping faucet can waste $200 to $500 worth of water in a year, and leaking toilets can waste several gallons an hour. Buildings and grounds departments, housing departments, and fellow students can encourage dormitory residents to act as stewards of their building and the environment and see that wasteful situations are remediated.

The keys to a successful stewardship program are training of building coordinators, prompt follow-up of reported leaks or other problems, and recognition and reward for taking the initiative to serve as a building steward. Most college dormitories have resident employees (students or older coordinators) who oversee student living, and most elect representatives to student government. University administrators can incorporate environmental stewardship training and responsibility into these existing structures. The effort can be organized to encourage and reward students to take responsibility for their living space. Stewardship initiatives can be coordinated with buildings and grounds departments so that repairs are made in a timely manner; failure to include this results in disillusionment with the program.

### Operate Buildings Efficiently

University housing staff and buildings and grounds staff have a responsibility to operate, maintain, and improve student housing. There is evidence that people are more likely to respect property that is well maintained. Also, requesting and expecting environmentally sound action from students is more effective if the building itself is operating efficiently. In the short run, the departments of student housing and buildings and grounds can work to ensure that storm windows are closed in the winter, heat and air-conditioning levels are appropriate, and recycling is well organized and running smoothly. In the longer term, effective programs include building design and construction.

Operating student housing within a reasonable and livable temperature range is a challenge on many campuses, particularly those with older buildings and inappropriate thermostat controls. Room temperatures can often vary widely within the same building, and correcting these temperature control problems can be costly. A university energy manager may find that temperature control is a top priority in some dormitories.

**Shut Down or Consolidate Residences During Vacations**
Operating a residence hall is resource intensive, and closing large halls during vacations can have significant savings. Students who are staying on campus during the holidays or vacation can be consolidated into a single dorm in order to reduce the need for security, lighting, heating, and water heating.

**Use Competition to Publicize Efforts and Increase Participation**
Competitions between dormitories have been a successful way to develop awareness for and measure the effectiveness of personal action to reduce waste. The competitions, originally called Eco-Olympics (a name that is now prohibited by the U.S. Olympic Committee) and now called Green Cup by National Wildlife's Campus Ecology Program, have been held at universities throughout the country.[4] The exact events that make up the competition vary from school to school but usually include monthly reductions of electricity use, increased collection of recyclables, and water conservation. Where they are quantifiable, measures of heating and trash generation might also be included.

At Tufts, students earned bonus points for using a reusable mug in take-out cafeterias, and at the University of Wisconsin at Madison extra points could be won for volunteering in the university recycling program and buying recycled products. Wisconsin students also worked with the university housing office with the goal of getting all 6,000 students involved in meeting a goal of saving $50,000 in utility costs and investing those savings in energy-saving technologies. Unfortunately this one-time contest had difficulty quantifying the results of the effort.

In some of its contests, Dartmouth measures progress on a per student basis, such as recycling per student or energy reduction per student, and the department of buildings and grounds has actually run the contest and measured the progress on their own meters. Prizes of pizza or ice cream were provided to winning dormitories. Appendix D describes a "decathalon" Tufts tried in 1992 as part of our Green Cup. Tufts CLEAN! estimated that the Tufts contest saved several thousand dollars in electricity and water, but found that organizing the contest was too labor intensive to run annually.

## Conclusions

Students are the university's unique resource—one that makes environmental stewardship innovative, exciting, and worthwhile. Often students are frustrated because they perceive environmental needs that faculty, staff, and administrators fail to address. However, there are countless ways that these same students can take leadership, in both their own behavior and choices, and act as valuable resources to the process of greening their campus. By combining their academic work with their volunteer activities and living experience, students can be leaders, participants, and resources for the university's environmental stewardship efforts.

# 9

# Greening the Ivory Tower

Where the smokestacks of a steel mill, the smell of a paper mill, or the round domes of a nuclear power plant command the public's environmental consciousness, a university's environmental impact may appear small by comparison. Even in a quiet college town, the college is rarely seen as a large polluter. Yet institutions of higher learning are polluters, contributing to local air, water, and land pollution. Universities generate radioactive, solid, and hazardous wastes; consume vast amounts of food; and purchase metals, paper products, fuel, water, and electricity that foul air, land, and water.

Universities and colleges should take their environmental impacts seriously and strive to meet, and then exceed, the regulatory requirements that govern them. It is the very nature of universities—as institutions that educate students for the future—that should motivate them to implement a vision of minimal impact on the natural world. The condition of the natural environment and the protection or restoration of this environment will influence a university's graduates in nearly every profession. By greening their own campuses, universities and colleges can teach and demonstrate the principles of awareness and stewardship of the natural world, while increasing their chances of clean and pleasant local and global environments for the future.

Greening a university is a long-term project that requires broad commitment and action. Careful and dedicated planning and organization, accompanied by statements of policy and conviction, can give any campus environmental stewardship effort a solid basis for implementation. This commitment and the resulting plans must then be incorporated into all university departments and functions, all while continuing to educate, feed, and house students and conducting faculty research. Taken as a series of unrelated actions, the list of tasks to green a campus is long and seemingly insurmountable. However, when these actions are part of an overall strategy and commitment to act, this same list becomes nearly routine.

On some campuses, the environmental stewardship effort begins with leadership from top administrators. In many places, however, the environmental stewardship effort will not have this top-level support, so initial steps must begin with and build on the participation of committed staff, faculty, and students in willing departments. In either case, the environmental consequences of nearly every decision—about purchases, operations, facilities, research, and even curriculum—should be considered and factored into decision making in order for the university to be truly green.

This chapter summarizes the characteristics of a university's culture that lead to environmental stewardship and presents ten lessons learned from the Tufts CLEAN! experience.

## A Culture of Stewardship

Failure to act or to think as a steward of our world is in fact an environmental choice. Most university decisions, such as to build a new dormitory or to purchase a new snowplow, are made with little thought to their environmental consequences, and yet the consequences over time can be significant in terms of energy use, wastes generated, and effect on local landscapes.

Changing the awareness and understanding of the ways in which everyday personal and institutional actions have long-term and cumulative consequences will be an important aspect of environmental action on campuses. But linking actions to their long-term and distant consequences is a difficult task. Even when the benefits or consequences of a choice are of a personal nature, such as the benefits of improving our individual diets and exercise habits, many of us find it difficult to make lasting change; it is no wonder that we fail to act when the benefits of our actions are less personal and tangible. As with the attempts that many individuals make to eat right, we often find that environmental actions are difficult because the choices are repetitious rather than a single change, and we tire of the need for self-restraint. The tools that reinforce environmental stewardship in our daily choices are the same tools used in personal lifestyle changes, such as quitting smoking or changing eating habits. Establishing a campus culture of stewardship—one that recognizes environmental problems and our connection to them and rewards choices with long-term benefits—can help to reinforce individual efforts and make the cumulative effect of the changes measurable.

A culture of stewardship reinforces and strengthens institutional efforts to incorporate efficiency and waste reduction directly into decisions and policies. This is particularly important when an institution is making large decisions, such as those concerning major construction and renovation of buildings and landscapes or long-term contracts with companies providing food service or pesticide applications. For example, a university wishing to build a new academic building should seriously consider renovating an old university building, since the resources used and the disruption to the land and water are likely to be less than the impacts from constructing a new building from the ground up. And if a new building *is* needed, the university should select a site with the smallest impact on surrounding water or lands.

In order for campus environmental initiatives to be far-reaching and eventually become institutionalized, top-level administrators must articulate a commitment to environmental action and demonstrate it in their own actions and decision making. A university environmental policy or presidential statement is useful for laying out the general principles of environmental stewardship and its importance. At the same time, a university administrator must recognize that his or her own office must take a leadership position in its actions, both large and small. For example, mailings from the administration can be printed double-sided, and fund-raising campaigns for new research facilities might include monies for environmentally sensitive technologies. The administration can create an expectation that environmental issues are being considered in all corners of the university and reward those who take action. When top-level administrators fail to understand this, as when they demand unnecessary luxuries or continue wasteful practices, they undermine efforts to reduce wastes that are underway throughout the institution. A president who takes the university's environmental vision seriously and demonstrates it by making recognizable changes in policies and personal habits, such as walking to work, will be followed.

The climate for environmental action at universities varies from school to school, and often from one campus to another within a large institution. The approach to reducing waste, conserving water, and using new environmental technologies on campus should mesh with the university's character and goals for education, action, and advocacy. To foster environmental stewardship, leaders should look at all aspects of the institution's character—its political disposition, attitude toward competition, religious underpinnings, or focus on health—and think about the individuals and groups that make up the distinct aspects of the institution's culture. Working within this culture, environmental stewardship efforts are likely to be more successful than if they run counter to the well-established culture.

As members of the university strive to build an environmental culture and address the campus and its impact on the environment, they may want to get outdoors—for meetings or strategy sessions perhaps—to see at firsthand the beauty and opportunities that are around them. Environmental leaders might take the provost or vice president on a visit around campus, walking among its plants, drinking its water, visiting its loading docks, and inspecting its hazardous waste collection sites. This tour might look in dumpsters, storm drains, chemical storage stockrooms, supply closets, and boiler rooms, asking if the grass is too green and the floors too clean for the health and safety of students, employees, and the surrounding community. Establishing a culture of environmental stewardship can be greatly facilitated by connecting the decision makers with the natural environment, the visible nature of the problems, and their solutions.

### When University Leaders Are Reluctant

In some institutions, a university-wide culture of environmental stewardship and leadership from top-level university administrators will be strong and visible; in many others, it will appear to be limited, ineffective, nonexistent, or even seem detrimental to campus environmental stewardship efforts.

When top-level commitment is limited, a university-wide committee or environmental leader should be certain that administrators understand the specifics of the desired commitment. Tufts CLEAN! found that managers or administrators did not always understand the problem or have enough information about what was asked of them. Individuals who seemed resistant were often simply overwhelmed by other priorities without obvious connections to environmental issues. Often, too, people are strongly invested in existing systems, technologies, and policies, because they meet the needs and may have resulted from considerable time and effort.

Top-level commitment and support are important components of any university environmental efforts, but their absence or weakness should not deter action. Campus environmental leaders are essential to implementing any initiatives, and they can lead important and far-reaching grass-roots efforts regardless of the support they receive from the university leaders. Many of these efforts, if successful, can be building blocks for broader commitment and buy-in. Each of the actions described in chapters 3 through 8 is a place for individual leaders to begin to decrease the environmental footprint of their college or university and build commitment in doing so.

In institutions that lack strong top-level commitment to environmental stewardship, it is important the environmental leaders try a cooperative and humble approach to working with administrators. Adversarial tactics such as public embarrassment, letter-writing campaigns, and sit-ins are not likely to foster the kind of cooperation that underlies long-term and successful efforts. Instead, students, faculty, and staff leaders should offer to investigate problems, invite administrators to join publicity events, and keep them informed of progress on projects even if those administrators were unaware that the project was beginning. If adversarial tactics are used, they should be used with care and a realization that they may break down trust and foster difficult working relationships.

### Expecting Environmental Action

In some companies, the job responsibilities of many personnel include attention to safety and environmental protection measures (such as wearing protective clothing and properly disposing of wastes). Requiring similar environmental action as part of job descriptions may or may not work in all campus operations, but it can be a model for some campuses, and one that makes environmental protection the norm rather than the exception. Environmental leaders must understand and acknowledge how implementing the principles of waste reduction, conservation, efficiency, and reducing life-cycle impacts can affect specific job responsibilities and expectations for performance. For example, grounds departments that embrace environmental stewardship may select plantings that are native to the region and withstand dry summer conditions rather than water-intensive plants that may be more colorful. Fostering environmental action should dovetail environmental efforts with legitimate university needs; for example, zealous efforts to turn out lights in order to conserve electricity cannot ignore the realities of security, nor should the collection of recyclables compromise fire safety.

Many university managers and administrators (as well as government officials and policymakers) feel that environmental stewardship stops at complying with laws governing air pollution, water discharges, and hazardous wastes. Nevertheless, simple compliance should be the beginning, not the end. Assurances that a university is in compliance with a particular law can sometimes fuel complacency and be a deterrent to more progressive action.

## Campus Learning for Environmental Stewardship

Campus environmental stewardship can be supported by several types of environmental learning. First, environmental leaders from the staff, faculty, and students learn about the latest energy-efficient and waste reduction technology and techniques. Second, members of the university community are offered opportunities to learn how they can reduce their impact on air, water, and land. Third, university curriculum may include the connection of each subject to environmental issues. The green university strives to undertake each of these types of environmental education.

Members of the university community who are looking to turn the concepts of efficiency and waste reduction into action must combine environmental literature with technical information on topics such as air handling systems, boiler maintenance, and food preparation. This literature provides important information about the nature of specific operational areas as well as proved, innovative, and environmentally beneficial solutions to targeted problems. Technical experts, rather than environmental advocates, are often the best sources for information.

Communication of information will be better if specific examples are used to illustrate general concepts. For example, an explanation of source reduction is strengthened when introduced in the context of a particular issue, such as the reduction of paper, and strategies to implement this objective, such as reduced mailings and double-sided copiers. Education can also occur by publicizing successful department initiatives and the environmental efforts of others. Dartmouth students used an innovative publicity stunt to illustrate the waste generated on campus; they personally hauled the trash they each generated in a week with them in knapsacks. Students at the University of Wisconsin developed an interpretive walk around campus to raise awareness. At Tufts, a slide show of environmental efforts throughout the campus was useful in publicizing existing actions and motivating progress. Outside the classroom, campus tours and admissions brochures can highlight environmental accomplishments or actions by featuring recycling areas and new CFC-free chillers. Like many other schools, Ohio State incorporates source reduction and recycling information into freshmen orientation. The University of Iowa's waste management coordinator has taught waste management courses for staff development and offers waste management approaches to empower those who must address the problems. The University of Colorado has multilingual posters and other written materials describing their recycling programs.

Within the classroom, environmental stewardship and action, rich in history, science, literature, and computation, can be woven into almost every course. The opportunities for hands-on and applied learning offer great benefits to student education.

## Lessons from the Field

Environmental action at colleges and universities is both similar to and different from efforts to reduce environmental impacts in homes, industries, hotels, and food service establishments. On the one hand, the issues—solid waste, hazardous materials, energy, water, and transportation—are similar, and the solutions require technology and methods similar to those used outside universities. On the other hand, the diverse goals and interests of the university community, as well as the university's educational mission, make the task difficult, diffuse, and seemingly distinct from the institution's priorities. The Tufts CLEAN! project sought to improve energy efficiency, conserve water, reduce and recycle wastes, and increase environmental awareness across the operations of three campuses (our main suburban campus, our medical school in downtown Boston, and our rural school of veterinary medicine). The project lasted four years and succeeded in developing an environmental policy statement for the university, establishing an environmental task force, and implementing countless environmental initiatives across the campus. Some of these initiatives succeeded, others have not, and still others have taken over four years to become fully implemented. In the process, the project learned some lessons that might help others to reduce their university's environmental footprint.

### Lesson 1: Environmental Stewardship Almost Always Means Reducing Waste

Almost every initiative to improve environmental quality will result in reducing waste on and off campus. For example, new lamps use less electricity than conventional lamps, and energy-efficient windows reduce heat losses in the winter and reduce the need for cooling in the summer. Debates about disposable cups are solved by selecting options that generate the least waste on the whole, such as lighter weight, smaller, less bulky, or reusable cups. Using microscale chemistry, fixing refrigerant leaks, and buying in bulk have the effect of reducing hazardous waste, CFC releases, and solid waste.

Waste reduction sometimes means financial savings, in either the purchasing costs or the longer-term operations. Cost savings are indeed important motivators for waste reduction, but it is important for environmental leaders to acknowledge that universities incur some expenses regardless of their financial payback. For example, universities hire staff or contractors to wash windows, paint field goal posts, and place leather chairs in the president's conference room. Cost savings are important measures of waste reduction but should not be the only measures.

**Lesson 2: Turn Concepts into Action**

Successful environmental stewardship relies on turning concepts into actions. Too often students and faculty conduct research that describes what needs to be done but falls short of laying out the details needed to get things done. It is generally easy to identify waste on a university campus and determine the conceptual actions for addressing these waste problems. Although these concepts are common and easily identified, connecting each with an appropriate course of action requires a detailed assessment of the situation in each location, finding funding, and coordinating with building managers, purchasing departments, building residents, and others. Many environmental audits stop short of asking and answering questions that are vital to the success of a project: Who will be responsible at each step of the process? Who will pay for the change? Who will oversee it? How will the office or department continue to operate while this change is undertaken? Environmental action must not become paralyzed by this attention to detail, but it rather must plan for the detail to avoid roadblocks. The question is how to undertake the series of actions that implement a single concept.

In a recent contest at Tufts, faculty, staff, and students were invited to enter their ideas for reducing waste and improving efficiency at the university. Winners received prizes of up to $500 in order to encourage detailed and thoughtful entries. The contest announcement requested that the applications include details about how the project would be carried out, by whom, the costs, and where the idea was already working. Only 1 of the 120 entries included information in sufficient detail to make any decisions about implementation. Environmental leaders who want to see action for environmental stewardship must be very specific about how to implement the program on their campus.

Implementing environmental action means focusing on scheduling, contracting with outside vendors, and dealing with the unique aspects of a university's facilities (e.g., low loading dock ceilings, inadequate wiring, or outdated plumbing). Project implementation may also include planning for education and information to change attitudes and behaviors. Evaluation and making modifications must also be included.

### Lesson 3: Environmental Solutions Go Beyond Technical Fixes

Solutions to university environmental problems require a combination of technology, institutional policy, and individual commitment and initiative. Some solutions may rely more heavily on one of these three types of solutions, but comprehensive programs require some of each. For example, an initiative to reduce campus electricity use encompass the installation of energy-efficient lamps, ballasts, and occupancy sensors. A university policy mandating the purchase of energy-efficient office equipment would further the goal. But to maximize the success of this program, individuals would have to follow the university's policy, choose the appropriate technology when buying new equipment, and turn off lights, computers, and other equipment when not in use. Technology-based solutions are advantageous because they make significant and dependable progress without relying on human choice, human memory, or individual inclination. Individual responsibility nevertheless is needed to implement and maximize the effect of both technology and policy.

As new technologies become available, new methods are discovered, or new students and employees join a university, there are fresh opportunities to make additional and incremental progress on environmental policies and problems. Because our world seems to become increasingly more complicated—more electronics, more packaging, more information—there is an increasing need for vigilance from the environmental stewards on and off our campuses.

### Lesson 4: The Largest Environmental Stewardship Opportunities Depend on University Operations

University staff in the facilities, grounds, food service, purchasing, and safety departments are essential to university environmental stewardship efforts. Enthusiastic students and faculty should seek out the pragmatic day-to-day wisdom of the university operations staff. In order to green Tufts, we found that we had to rely on the operations staff, who could make changes to the physical plant, buy new materials, or change the menus in the dining halls. Rather than undertake initiatives independently, we supported these staff members with time, information, and expertise. In addition, we sought the connections among the environmental efforts of various departments in order to help institutionalize and build the effort.

Working with the university staff requires a respect for the ways in which the performances of operations division staff are rewarded and an acknowledgment of their primary responsibilities. For example, dining managers must always put sanitation ahead of food waste composting, and purchasing managers must pay close attention to cost. Environmental action will be most successful when university staff responsibilities reflect and reward attention to waste reduction, recycling, efficiency, and safety in the same ways that job performance rewards attention to these more immediate and familiar issues.

**Lesson 5: Environmental Issues Can Link the Resources of Students, Staff, Faculty, and Administrators**

Much study of and research on environmental problems is focused on distant places, such as the tropical rain forest, the northern tundra, and the depths of the sea, and in the process, universities often overlook the problems on their own campuses. During the Tufts CLEAN! project, we were struck by the ability of the university community to solve problems creatively when we linked the innovation and theoretical thinking of faculty, the pragmatism of staff, and the energy and idealism of students. This combination of theory, practice, and vision is a powerful and often underused combination for far-reaching university environmental action. When faculty, staff, and students work cooperatively with each other, they increase their understanding of and respect for the needs of each other and are thereby less likely to miss opportunities to use the university's own campus as a testing ground for new ideas and devices.

Campus environmental issues provide opportunities for everyone on campus to be involved as participants, investigators, and intellectuals. For example, solid and hazardous waste issues are rich in history and often have direct connection to issues of justice and gender. Questions such as where the university's solid waste goes and how radioactive material is stored on campus may be of interest to political scientists and sociologists. Physics students can study the optical properties of the fluids in underground storage tanks as their levels change. Computer programmers might undertake a project to rid the alumni mailing list of duplicate names, or they might study ways of remotely monitoring building heating and air-conditioning.

**Lesson 6: Audit for a Purpose**

At many U.S. universities, students and faculty are undertaking broad-based environmental audits of their campuses, collecting and compiling detailed data about resources used and wastes generated. The resulting audit is useful for raising awareness but is often too general to inform decisions. General recommendations such as "install water-saving showerheads," "install energy-efficient lighting," "recycle white paper," and "implement a bike-to-work program" are common in the audit. However, these same recommendations can also be created from a simple walk around campus and a reading of basic books about conservation. At Tufts, we spent the better part of a year collecting aggregate information about the university's environmental impacts, finally realizing that detailed solutions required more than information about aggregate impacts. Our data collection efforts also taxed staff, whose time was needed to work on solutions.

Auditing, the first step in environmental action in companies, is very complicated and highly variable at universities because the inputs and outputs of producing the products—educated students and research results—are complicated and variable too. In universities, the audit process can be used successfully to gather detailed data that inform direct and specific action. For example, collecting data on the total electricity used in an individual building is not helpful in shaping a strategy to reduce electricity usage in that building. On the other hand, a detailed audit of the existing lamps and ballasts and existing light levels in each room of that same building is essential for determining the retrofits that are required.

**Lesson 7: Goal Setting Is Powerful But Difficult**

Setting goals can help to motivate, focus, and prioritize a university's environmental efforts. Goals help to translate the spirit of environmentalism into measurable results. However, setting appropriate goals can be difficult because university faculty, staff, students, and environmental leaders lack good data and a sense of what is possible. Furthermore, goals that reflect progress in some important areas, such as the awareness of the university community, are difficult to quantify and measure.

Rather than spend a great deal of time conducting analysis to inform goals, universities might consider taking advantage of the goals that already exist in international, national, state, and local policy. For example, if the local government seeks to reduce water use by 5 percent in order to meet the demands for water created by new homes, then the university must also reduce its own water use by at least 5 percent, or other water users will have to make up the difference. Or a university could commit to eliminate the use of CFCs on campus prior to the products' phase-out under current law or pledge to hold emissions of carbon dioxide at 1990 levels by the year 2000 as required by agreements signed by many nations at the 1993 World-wide Conference on the Environment held in Rio. Table 9.1 lists examples of goals that are appropriate targets for universities by providing measurable and achievable milestones.

**Table 9.1**
Appropriate goals for university environmental action

| Goal | Source/originator | Ways to make progress toward this goal |
|---|---|---|
| Elimination of CFCs by 1994 | Federal law: Clean Air of 1990 | Change to alternative refrigerants |
| Stabilize carbon dioxide emissions at 1990 levels by the year 2000 | International conference and treaty: Rio Conference 1993 | Hold electricity use at 1990 levels by improving the efficiency of lights |
| Reduce hazard emissions by 33 percent from 1988 levels by 1992 and by 50 percent by 1995 | Federal program: EPA's voluntary 33/50 program | Switch to microscale techniques for teaching chemistry; clean out old labs |
| Recycle 35 percent of solid waste | State recycling goals | Recycling paper, cans, cardboard, plastics |
| Reduce traffic and congestion | Town or city priority | Carpooling program |

A goal of no net additional impact is appropriate for institutions that plan to be in business well into the future. Meeting a goal of no net impact (in either absolute or per student terms) will require that the university offset any gains in electricity demand, water use, and solid and hazardous waste generation with corresponding improvements in efficiency and conservation. For example, Oberlin College's Environmental Studies Center plans to be a net exporter of energy. A no-net-gain policy will provide the university with insurance against increases in the costs of power, water, wastewater treatment, and waste disposal. In most universities, a no-net-gain policy is cost-effective. As universities become more efficient and less wasteful but continue to expand facilities or programs, finding ways to offset campus expansion will become more and more difficult to achieve.

### Lesson 8: Take Action Where You Can Be Successful

The university's larger environmental problems may require complicated or far-reaching solutions, many players, changes in infrastructure and systems, or capital. Often campus environmental leaders may be able to make the most progress on smaller projects, where the costs are low, players eager, and solutions easy. At Tufts, we found we could save more water by undertaking a "turn off the faucet" campaign in dining halls than we could by installing a sprinkler system to irrigate the athletic fields, despite the far greater water saving potential with the irrigation systems. This resulted because we were welcome in the dining halls, but the grounds department had other priorities and was reluctant to change.[1]

Situations when health and safety are immediate concerns or laws or permits are being violated should be considered top priorities. Working with university staff to fix these problems and reduce or avoid liability and fines will be in everyone's interest.

### Lesson 9: Seek Institutionalization Rather Than One-Time Events

The formal Tufts CLEAN! project lasted as long as its outside funding. In those four years, we identified many problems and solutions, worked closely with some departments, and implemented numerous changes. Perhaps most important, we were a constant and persistent voice for change, providing information and assistance to the university. Except in rare cases, we did not actually carry out the changes ourselves; instead we relied on dining services managers, purchasing managers, ground crews, building managers, and countless others. Often our team felt that progress was too slow and fell short of our comprehensive vision.

Now, nearly three years later, there is strong evidence that environmental stewardship is alive and well in many operational departments and that environmental stewardship is part of many of Tufts' everyday activities. The operations department has hired a full-time energy manager, the custodial staff now considers recycling as part of their responsibilities, dining services is trial-marketing organic foods, purchasing is pushing recycled paper, and the university will soon reconvene its environmental committee.

Institutionalization means that the changes are likely to last longer than one-time, high-publicity events. It also often means that change may take longer to achieve. Institutionalization requires taking the time to build systems—committees, policies, and training—that will support far-reaching and long-lasting change. Building environmental knowledge and action into the jobs of the university's faculty, staff, and students will support ongoing and sustained action.

One strategy for institutionalization of a university environmental policy is to ask departments across the university to report to the president, the administration, or the environmental committee on their own goals for implementing the environmental policy in their own departments and the progress they are making toward implementing these goals.

**Lesson 10: Never Take No for an Answer**

Achieving the comprehensive environmental stewardship described throughout this book will not be easy, even at the most dedicated university or college. Some institutions have gone further than others in implementing change, but along the way there have been and continue to be barriers to that change, regardless of the place.

Many of these barriers can be overcome by changing the approach to the problem. In most instances, colleges and universities tend to operate buildings, maintain grounds, and make decisions about purchasing or travel in ways consistent with past habits and procedures. In part, universities conduct business and affairs in this way because the status quo has evolved to solve a set of problems or conditions, and in part because it is an easier and more comfortable way to proceed. Changing habits and procedures requires diligence, willingness to take risks, and patience when efforts go awry. In effect, environmental stewardship involves reinventing the way that the university's departments approach decisions and everyday procedures.

Barriers to environmental stewardship can also be overcome by looking outside the university for expertise, motivation, and funding. Environmental leaders and operations managers may find that college alumni offer a wealth of knowledge about the solutions to environmental problems. Other institutions and organizations are likely to have relevant experience on almost any topic (see Appendix E for relevant Internet resources). Funding may also be obtained from alumni as well as from corporations and foundations. Although faculty are familiar with writing proposals to solicit funding, operations managers are not, so the partnerships of an environmental committee can be instrumental in this case. Box 9.1 shows some barriers to campus environmental stewardship that are often raised by campus environmental leaders and proposes some actions to overcome them.

## The Green University

The green university of tomorrow uses resources with care, considers the environmental impact of all its decisions, and makes these decisions with an eye toward waste reduction, energy efficiency, and reducing life-cycle impacts. It uses its environmental stewardship ethic to guide teaching, research, and operations. This university's environmental footprint does not grow over time, since each new impact is offset by efficiency and waste reduction improvements elsewhere on campus. The university uses its investments to promote other environmentally sensitive initiatives and incorporates environmental teaching into all courses. Fiscally and academically, this university is sound, and it will continue to serve its students and alumni for years to come. David Orr argues that

> the crisis of the biosphere is symptomatic of a prior crisis of mind, perception, and heart. It is not so much a problem in education but a problem of education. . . . It is rather about basic and mostly unstated assumptions that affect the ecological literacy and competence of students, faculty, and those charged with admin-

**Box 9.1**
Barriers to campus environmental stewardship

*Problem:* Our administration is skeptical and lacks commitment.

**Things to try:**
• Point out that other universities are demonstrating commitment. Over 200 presidents and rectors have signed the Talloires Declaration (see appendix A).
• Demonstrate projects that have benefit to the university in areas they understand: reputation, enrollment, and dollars.
• Put words in their mouth. A seeming lack of commitment may just be that there are other priorities or a crisis that needs attention.
• Invite them to an environmental committee meeting.
• Don't wait for top commitment if you don't have it. There is a lot that can be done without it.

*Problem:* We are having trouble turning our broad statement of commitment into action.

**Things to try:**
• Break the action you want to achieve into manageable pieces
• Focus on a building, department, or section of the campus rather than on an issue (e.g., energy).
• Use small working groups.
• Develop concrete actions rather than concepts. Ask who will implement it? How? For how much?

*Problem:* We can't set priorities from among the many opportunities we have identified.

**Things to try:**
• Focus your effort where there is opportunity: where there is willing staff, where the environmental objectives dovetail with other priorities, or where you have expertise.
• Select a project because it will succeed. This will give you momentum for other projects.
• Don't expect a general campus audit to help set priorities. It is difficult to set priorities when comparing tons of solid waste to kilowatts of electricity.
• Select a state, national, or local goal, and pledge to do your part as a university.

*Problem:* Our staff lacks interest and/or commitment.

**Things to try:**
• It is unlikely that the entire staff lacks interest. Find the staff members who are interested and work with them.
• Work hand in hand with staff. They are the key to most projects.
• Publicize the actions of staff. Give credit generously to empower people. Use the newspaper, tours, radio, and slide shows.

**Box 9.1 continued**

• Show staff how some of their actions already have environmental benefit. Give people opportunities to tell you what they are already doing.
• Ask for ideas about how to reduce environmental impacts rather than telling people what to do. Prepare for meetings and get people to tell you what you want to hear rather than talking to them.
• Recognize and respect the demands on staff time (getting meals served, etc.).

*Problem:* Many of the students are apathetic.

**Things to try:**
• Lead by example.
• Don't try to drag people. Make change easy for them.
• Be specific about the actions you want.
• Find students and student groups with similar priorities. For example, the environmental club can join forces with the community service club.

*Problem:* Our campus audit is taking a long time.

**Things to try:**
• Make sure you know why you are auditing. How will the information be useful? If the audit will produce only general recommendations (e.g., reuse paper, turn off lights, carpool), you need a new approach.
• Realize that the general campus audit may not help set priorities.
• Target your audit efforts. For example, rather than gathering data on total water use, audit the shower heads in a single dorm to determine their flow rate. Find alternatives, and calculate the costs (including installation using university-approved labor) and the savings.
• Offer your audit team to the buildings and grounds, purchasing, or dining staff. Track down the things they are interested in. These will be much narrower projects, but are more likely to get implemented than general recommendations.

*Problem:* Some of our efforts or programs start with a bang but then fizzle.

**Things to try:**
• Don't worry. The fizzle may be a sign that the time is not right, or the solution you have chosen is not the right one.
• Focus on the actions that can work. Find the right people to champion them.
• Stay committed. This is hard work.
• Be action oriented. Move beyond concepts and be concrete. Who? How? How much?
• If you can get the provost or president or a dean to push you, do it. Ask him or her to set some deadlines for you or your committee.

**Box 9.1 continued**

*Problem:* We need to educate the university community (e.g., about recycling).

**Things to try:**
• Education needs to be ongoing. Establish a regular column in the student and/or staff paper.
• Use existing vehicles such as student or staff newspapers or newsletters, radio or TV shows, and the Internet.
• Make education an integral and ongoing part of any program.
• Student groups can conduct awareness campaigns.

*Problem:* It seems there is never any money to implement things.

**Things to try:**
• Money and time are always in short supply. Try to link environmental projects with other priorities.
• Try to capture savings realized from avoided costs in order to finance subsequent projects.
• Use student projects to research options.
• Use the campus to demonstrate new technology developed by faculty members.
• Spread the work—try to find appropriate places for environmental responsibilities. Fold environmental responsibility into job responsibilities.

istration and oversight. Ultimately, the crisis of the biosphere is a test of our commitments. . . . Colleges and universities must learn to act responsibly in such matters not only because it is right to be responsible, but also because it is in their self-interest.[2]

In order to achieve this vision today, we need to change the thinking behind routine decisions of university administrators, faculty, staff, and students. This will take time, and the extent of what is possible and realistic will change over time as costs rise, technologies change, and awareness and understanding increase. Both the individual, in his or her own work, and the institution, in its policies and expectations, have a role to play in achieving this vision.

Although the biggest opportunities for creating change largely fall in the domain of operations in departments like facilities management, dining services, purchasing, and environmental health and safety, collectively the students, academic departments, and administrative offices also have the potential to make a marked impact. Because universities and colleges are teaching and research institutions, their efforts, both large and small, extend far beyond the institution's walls. Those members of the university community who take the time and make the choices will be rewarded, if only with personal satisfaction and a knowledge that they are making a difference.

The enormous number and complexity of actions that occur in the green university mean that the process of greening the university is an incremental one that builds on itself and involves all members of the community. Along the way there are many opportunities for discovery, innovation, and creativity. Wendell Berry wrote: "The real work of planet-saving will be small, humble, and humbling, and (insofar as it involves love) pleasing and rewarding. Its jobs will be too many to count, too many to report, too many to be publicly noticed or rewarded, too small to make anyone rich or famous."[3]

Those of us who are members of university communities, as faculty, administrators, staff, students, and alumni, must persevere, taking the steps, large and small, as stewards of our natural world.

# Appendix A: The Talloires Declaration

*The Talloires Declaration resulted from a meeting of thirty-five university presidents who convened to discuss universities' roles in environmental protection and education. The 1990 meeting was held at Tufts University's campus in Talloires, France. Since then, more than 250 institutions in over 40 countries have signed this declaration. The declaration is now supported by the University Leaders for a Sustainable Future (ULSF), which offers memberships and member support services.*

We, the presidents, rectors, and vice chancellors of universities from all regions of the world are deeply concerned about the unprecedented scale and speed of environmental pollution and degradation, and the depletion of natural resources. Local, regional, and global air and water pollution; accumulation and distribution of toxic wastes; destruction and depletion of forests, soil, and water; depletion of the ozone layer and emission of "green-house" gases threaten the survival of humans and thousands of other living species, the integrity of the earth and its biodiversity, the security of nations, and the heritage of future generations. These environmental changes are caused by inequitable and unsustainable production and consumption patterns that aggravate poverty in many regions of the world.

We believe that urgent actions are needed to address these fundamental problems and reverse the trends. Stabilization of human population, adoption of environmentally sound industrial and agricultural technologies, reforestation, and ecological restoration are crucial elements in creating an equitable and sustainable future for all humankind in harmony with nature. Universities have a major role in education, research, policy formation, and information exchange necessary to make these goals possible.

University heads must provide the leadership and support to mobilize internal and external resources so that their institutions respond to this urgent challenge. We, therefore, agree to take the following actions:

1. Use every opportunity to raise public, government, industry, foundation, and university awareness by publicly addressing the urgent need to move toward an environmentally sustainable future.

2. Encourage all universities to engage in education, research, policy formation, and information exchange on population, environment, and development to move toward a sustainable future.
3. Establish programs to produce expertise in environmental management, sustainable economic development, population, and related fields to ensure that university graduates are environmentally literate and responsible citizens.
4. Create programs to develop the capability of university faculty to teach environmental literacy to all undergraduate, graduate, and professional school students.
5. Set an example of environmental responsibility by establishing programs of resource conservation, recycling, and waste reduction at the universities.
6. Encourage the involvement of government (at all levels), foundations, and industry in supporting university research, education, policy formation, and information exchange in environmentally sustainable development. Expand work with nongovernmental organizations to assist in finding solutions to environmental problems.
7. Convene school deans and environmental practitioners to develop research, policy, information exchange programs, and curricula for an environmentally sustainable future.
8. Establish partnerships with primary and secondary schools to help develop the capability of their faculty to teach about population, environment, and sustainable development issues.
9. Work with the U.N. Conference on Environment and Development, the U.N. Environment Programme, and other national and international organizations to promote a worldwide university effort toward a sustainable future.
10. Establish a steering committee and a secretariat to continue this momentum and inform and support each other's efforts in carrying out this declaration.

For further information about the Talloires Declaration or to sign the declaration contact:

University Leaders for a Sustainable Future
c/o Center for Respect of Life and Environment
2100 L St., NW
Washington, DC 20037
phone: 202/778-6133
fax: 202/778-6138
e-mail: ULSF@aol.com
web site: www.ulsf.org

# Appendix B: Tufts Dining Strategic Plan: Environmental Stewardship Initiatives

Created by Tufts Dining Environmental Committee

## Statement of Purpose

Dining Services needs to expand existing initiatives by developing procedures to reduce or eliminate adverse environmental impacts.

Dining Services is committed to incorporating the guidelines of the Tufts Environmental Policy in its operations by educating customers and staff in reevaluating operating procedures to offset the adverse environmental impacts of our industry.

We seek to play a leadership role in our industry and on campus. We will support other campus environmental initiatives.

**Key Result 1**
Reduce the toxicity and volume of wastes

| Strategies | Outcomes | Tactics |
|---|---|---|
| Examine alternatives to toxic materials and pesticide use. Criteria should include biodegradability, frequency of use, reduced toxicity and concentration. | Dining uses materials with lowest toxicity. | 1. Select five areas to focus on. Include concentration, pesticides, and cleaning chemicals. 2. Examine toxicity of these products. 3. Identify and test alternatives for problem products. 4. Identify and implement employee training or changes in procedures. 5. Periodically evaluate effectiveness and repeat for other products as needed. |

**Key Result 1** ***continued***

| Strategies | Outcomes | Tactics |
| --- | --- | --- |
| Take steps to reduce and recycle packaging of food and supplies delivered. | Deliveries use least packaging possible. | 1. Develop criteria of preferences for packaging, including reusable and recyclable products.<br>2. Write letter to internal Tufts food purchasing staff to discourage things such as retail packaging.<br>3. Write letter to outside vendors.<br>4. Add appropriate criteria to bidding specs.<br>5. Develop audit questions to target action.<br>6. Expand existing recycling. |
| Reduce packaging in which food is served. | Per meal packaging is minimized. | 1. Use reusables wherever possible.<br>2. Conduct customer awareness campaign to reduce loss (theft).<br>3. Determine and rank alternatives.<br>4. Select and test alternatives.<br>5. Conduct customer education.<br>6. Determine if selected alternatives are pleasing to customers. |
| Future facility renovations should include opportunities for source reduction. | Future facilities are not a barrier to environmental action. | 1. Incorporate appropriately sized dish-washing facilities and storage.<br>2. Properly design waste disposal areas to incorporate source separation. |
| Reduce amounts of pre- and postconsumer food waste. | Little or no food waste is disposed of as trash. | 1. Examine opportunities for preprocessed foods.<br>2. Educate employees.<br>3. Educate customers. |

**Key Result 2**
Improve energy efficiency and water conservation

| Strategies | Outcomes | Tactics |
|---|---|---|
| Review and improve day-to-day practices seeking opportunities for efficiencies and conservation. | 1. Lower energy and water costs.<br>2. Recommendations of improvements are developed.<br>3. Greater employee understanding. | 1. Educate employees by providing examples, sharing real costs, and informing them of environmental impacts.<br>2. Review/audit each facility with managers. (a) Develop an audit; (b) Select audit captain.<br>3. Develop a way to get employee input (e.g., contests).<br>4. Implement changes and provide feedback where possible.<br>5. Develop ongoing checking mechanisms. |
| Retrofit and upgrade equipment. | Decreased energy and water consumption by dining equipment. | 1. Develop guidelines for efficiency and clean fuels.<br>2. Identify energy (fuel) characteristics of equipment soon to be replaced and determine alternatives and select new equipment for efficiency.<br>3. Inventory existing equipment for energy needs and prioritize.<br>4. Replace/retrofit equipment in order of priority.<br>5. Determine if selected alternatives are pleasing to customers. |
| Future facilities renovations should consider efficiency and conservation in the design phase. | Future facilities are as efficient as possible. Per meal utility use decreases. | 1. Use Tufts design standards, or develop those suited to dining.<br>2. Make standards available to architects. |

**Key Result 2** ***continued***

| Strategies | Outcomes | Tactics |
|---|---|---|
| | | 3. Include standards as objectives of design.<br>4. Include people with expertise in planning process.<br>5. Examine possible use and funding opportunities for renewables. |

**Key Result 3**
Reduce adverse environmental and health impacts of food choices

| Strategies | Outcomes | Tactics |
|---|---|---|
| Examine food sources to assess opportunities to reduce transportation and health risks (pesticides and hormones). | Foods come from more local sources, and use of organic or low-spray alternatives is maximized. | 1. Research locally available items (student project) such as honey, potatoes, turkey, and root vegetables.<br>2. Match Tufts' needs, availability, and cost.<br>3. Identify applicable health risks due to processing, pesticides, and hormones.<br>4. Modify menu to reflect seasonally available local products.<br>5. Target alternatives to reduce concerns.<br>6. Provide customer education.<br>7. Identify ways to balance additional costs against associated savings. |
| Provide menu alternatives that are lower on the food chain. | Alternative foods are readily available. | 1. Expand vegetarian meal options.<br>2. Extend Healthy Balance options [a Tufts program to increase |

**Key Result 3**
Reduce adverse environmental and health impacts of food choices

| Strategies | Outcomes | Tactics |
|---|---|---|
| | | vegetarian and low-fat foods] to other facilities.<br>3. Offer special vegetarian nights (use in-season vegetables) featuring organic grains and intercultural recipes.<br>4. Provide customer education re: environment, health, cost to encourage people to try vegetarian meals.<br>5. Offer vegetarian meal plan at lower cost.<br>6. Develop regular way of involving purchasing, students, and dietitian. |

**Key Result 4**
Increase customer and staff awareness of environmental issues

| Strategies | Outcomes | Tactics |
|---|---|---|
| Utilize available media to educate and promote dining environmental programs. | Programs are better known. | 1. Use dining menus, call-in line, monthly newsletter to promote environmental issues.<br>2. Use Tufts publications to generate publicity and awareness.<br>3. Use bulletin boards, table tents, TUTV, and TU radio.<br>4. Hold a Cause Night with vegetarian meals, and proceeds to go to environmental groups.<br>5. Organize message by themes. |

| **Key Result 4 continued** | | |
|---|---|---|
| Strategies | Outcomes | Tactics |
| Conduct staff training. | Staff are aware of environmental actions associated with performing their jobs. | 1. Create comprehensive training.<br>2. Test and evaluate in one location.<br>3. Conduct training in all halls.<br>4. Include environmental issues in annual Employee Day. |
| Work with student groups. | Student input is sought and influences environmental decisions. | 1. Appoint student representatives to Dining Environmental Committee.<br>2. Solicit input from Student Dining Concerns Committee.<br>3. Cosponsor events with students. |
| Take steps to play a leadership role in the food service industry. | Tufts is viewed as a leader in environmental stewardship among university food service departments. | 1. Publicize results via NACUFS [National Association of College and University Food Services].<br>2. Hold formal/informal meetings with area colleges to discuss issues.<br>3. Go beyond industry forums. |

# Appendix C: Dining Services Assessment Worksheet

| Equipment | | | Operation | | | Opportunities |
|---|---|---|---|---|---|---|
| Description and location (e.g., oven, toaster, etc.) | Type (gas, electric) and manufacturer | Condition (new, clean, dirty, leak-ing, etc.) | Meals | Hours including vacations (e.g., 6 am–9 pm; always even during vacation) | Levels (always on high, etc.?) | Efficiency, repair, replace, shut down, etc. |
| | | | | | | |
| | | | | | | |
| | | | | | | |
| | | | | | | |
| | | | | | | |
| | | | | | | |
| | | | | | | |

# Appendix D: Tufts Eco Cup Decathlon

Created by Karen White and Peter Allison, Tufts University, 1992

## A Ten-Stage Resource Conservation Bonus Event

The Tufts Eco Cup Decathlon is a ten-stage event that pits hall against hall in a battle to identify NEW WAYS to conserve energy and resources. Unlike most other Tufts Eco Cup events, which challenge students to test their own limits (e.g., by taking shorter showers), in the Decathlon, students tell the university to go for the gold by making investments that make *cents*.

For each of the ten stages, your hall can win a gold, silver, or bronze medal, and get 5, 3, or 1 points in the Tufts Eco Cup contest. The ten stages are described below. Points are awarded on how thoroughly you seek to find problems and suggest solutions for energy and resource conservation. With a total possible score of **50**, you could get enough points to streak to victory in the Tufts Eco Cup!

Decathlon scorecards will be provided to the Tufts Eco Cup representative for each hall and will also be available in the Residential Life offices. A hall can submit as many scorecards as they wish, but the total score for each stage cannot exceed 5. Scorecards must be dropped off at Residential Life (uphill in Carmichael or downhill in Hodgdon) by **Friday, March 13,** to count toward Tufts Eco Cup.

If you have any questions or need tips on how to proceed, please call Karen White at 627-3704 or Peter Allison at 627-3486. Again, thanks for competing—sorry we can't provide whirlpools.

## The Ten Stages

1. The chill of victory

Many window and doors on campus are poorly insulated and leak a lot of cold air in and warm air out. To earn points for this stage, identify windows and doors that can be better insulated. One way to get this information is to go door to door and ask people. This is also a good way to tell people about the Tufts Eco Cup contest. Another strategy is to post a sign in the bathroom stalls asking people to list their rooms.

2. The agony of de-*heat*
Are rooms in your building so overheated that people open windows on cold days to compensate? In some buildings, it is hard to control the heat because thermostats and temperature controls don't work properly. To deal with this problem, we need an accurate accounting of rooms that are too hot. To earn points for this stage, document the **rooms by number** that are usually overheated.
3. Sweating it out?
Another reason that some rooms are hot is that pipes are not insulated. To earn points, make a list of places where you find hot, uninsulated pipes running through your building. Once again, it may be useful to go door to door, as many of these pipes may be found in individual rooms.
4. Bowling for watts!
*Strike* out against unnecessary lighting and *spare* valuable resources. Many lights are left on in common areas even when nobody is there. Occupancy sensors are devices that detect motion and turn lights on and off when people enter and exit a room. To earn points for this stage, give us your ideas on what lights in which rooms could use an occupancy sensor. Lounges and study rooms are a good place to start.
5. Let the sun light the torch
Are there areas in your hall that are well lit by the sun during the day **and** lit by electric lights? These are places where a photosensor might mean big energy savings. Photosensors automatically turn lights off when a certain amount of natural light is in a room. To earn points for this stage, scout out and list areas that could use photosensors in your building.
6. Leak it to Beaver
Even if you turn off the water while brushing your teeth, if the bathroom faucet leaks all day, your hall's water usage will increase. To earn points for this stage, tour your hall and identify all the leaking faucets and dripping showers. If a fixture is dripping, test to see if it's just not turned off completely. Document faucets that leak or are hard to turn off.
7. Toilet (over)training
Do your toilets continue to run after flushing? If so, a lot of water is being wasted. To earn points for this stage, write down all the toilets that run too long.
8. Hit the showers
Many of the showers in the halls have low-flow shower heads, which save an enormous amount of water—yet there are also many that have outdated water-guzzling heads. To earn points for this stage, measure the rate at which water flows from showers in your hall (how long does it take to generate one gallon of water when the shower is on full? Do this while showering to save water!) Use a one-gallon milk jug, and a watch with a second hand. Document which showers in which bathrooms you measure.
9. Sorting things out
It is important to know what is in our trash in order to plan source reduction and recycling strategies. To earn points for this stage, sort through your hall's

trash cans and note those materials that are taking up the most volume and weight. Use a box and bathroom scale to weigh each component of the waste, and estimate the percentage of volume each material represents. For a gold you must conduct 5 sorts, for a silver 3, and for a bronze 1. See score sheet for details.
10. Free-style
Our list of opportunities for making the residence halls more environmentally sound is far from complete. Win points for this last stage by coming up with your own ideas on where we could make improvements. Be creative! Even off-the-wall ideas have value in decision making.

## SCORECARD: Tufts Eco Cup Decathlon

Residential hall: ______________________________

Names and phone numbers of ECOlympians:

______________________________________________

______________________________________________

______________________________________________

______________________________________________

1. The chill of victory
Identify room windows and doors that can be better insulated.

2. The agony of de-*heat*
List rooms that are overheated.

3. Sweating it out?
Identify rooms with uninsulated pipes.

4. Bowling for watts!
List locations for occupancy sensors.

5. Let the sun light the torch
List locations for photosensors.

6. Leak it to Beaver
Identify leaking faucets, dripping showers, and hard-to-turn nozzles.

7. Toilet (over)training
List toilets that run too long.

8. Hit the showers
List the flow of showers in your hall. Be specific about which floor, bathroom, and showerheads you measure.

| Floor | Bathroom (e.g., men, women, east) | Showerhead (draw a picture if needed) |
| --- | --- | --- |

9. Sorting things out
Date(s)
Source(s) of waste: Floor ______________ Room or hallway ______________

| | Wt %vol | Wt %vol | Wt %vol |
| --- | --- | --- | --- |
| Newsprint | | | |
| White paper | | | |
| Colored paper | | | |
| Food packaging | | | |
| Deposit bottles and cans | | | |
| Nondeposit glass | | | |
| Other | | | |
| Total | | | |

10. Free-style
Other suggestions. Feel free to attach other sheets to describe other ideas for reducing environmental impacts in the residential halls.

# Appendix E: Internet Resources for Campus Environmental Efforts

The World Wide Web offers hundreds of resources for college and university environmental efforts. Some of the most relevant and helpful are listed here. Sites for individual college and university efforts are not listed, but many can be found in the notes of this book.

## Mailing Lists

### Green Schools

The GRNSCH-L list is a mailing list for the discussion of university and college efforts to reduce environmental impacts. Send an e-mail message to listserv@brownvm.brown.edu to subscribe. The body of the message should say subscribe grnsch-l your name. The Web site is http://www.envstudies.brown.edu/cgi-bin/lwgate/grnsch-l.

### RECYC-L

This is a mailing list dedicated to the discussion of recycling programs at college and university campuses and related topics of interest to persons responsible for planning and implementing such programs. Send a message with the subject "SUB REQUEST" to recyc-l@brownvm.brown.edu to subscribe. In your message, include a brief introduction about yourself and your program. This message is sent to the entire list so all RECYC-Lers know who is joining the list. Your address will then be manually added by one of the List-Owners.

### Environmental Studies List Hosted by Brown University

This is a moderately active group discussing a variety of environmental issues often relating to education and teaching. Send an e-mail message to listserv@brownvm.brown.edu to subscribe. The body of the message should say: subscribe envst-l your name.

## Organizations and Information Resources

http://www.nwf.org/campus
The Campus Ecology Program at the National Wildlife Federation offers resources and networking for students, faculty, and staff interested in reducing environmental impacts on campuses. This organization can provide targeted resource materials and regional coordinators to assist campuses.

http://earthsystems.org/curc.html
This site provides information about the College and University Recycling Coalition of the National Recycling Coalition.

http://www.epa.gov
The U.S. Environmental Protection Agency offers numerous technical and regulatory publications and other information.

http://www.ccar-greenlink.org
GreenLink provides information about the environmental issues associated with maintaining vehicles (e.g., used oil, air-conditioning).

http://www.brown.edu/Departments/Brown_Is_Green/
This site provides an overview of Brown University's environmental efforts and provides links to many college and university campus-based environmental action sites.

http://www.cfe.cornell.edu/wmi
The Cornell Waste Management Institute serves the public through research, outreach, training, and technical assistance programs in solid waste disposal, management, and planning.

http://www.seac.org
The Student Environmental Action Coalition is a student-run and student-led national network of progressive organizations and individuals whose aim is to uproot environmental injustices through action and education.

http://www.greenseal.org
Greenseal is an independent nonprofit organization dedicated to helping people and institutions buy products and services that are less damaging to the environment. Their scientists study and test products to identify those that are friendliest to the plant and award the best products with a seal of approval.

http://greenbuilding.ca/GBIC.html
The Green Building Information Council is a Canadian nonprofit organization whose mission is to disseminate information about energy and environmental issues in the building sector. GBIC is a small organization with extensive interna-

tional links, and through this site it brings a broad range of information from other sources and organizations around the world.

http://www.usgbc.org
The U.S. Green Building Council is a nonprofit, consensus coalition of the building industry, promoting the understanding, development, and accelerated implementation of Green Building policies, programs, technologies, standards, and design practices. The organization hosts an annual conference.

http://www.solstice.crest.org
Solstice is the Internet information service of the Center for Renewable Energy and Sustainable Technology (CREST).

http://www.rmi.org
The Rocky Mountain Institute is a nonprofit research and educational foundation with a vision across boundaries. Its mission is to foster the efficient and sustainable use of resources as a path to global security.

http://www.appa.org
This is the Web site for the Association of Higher Education Facilities Officers.

http://www.nacufs.org
This is the Web site for the National Association of College and University Food Services.

http://www.nacubo.org
This is the Web site for the National Association of College and University Business Officers.

http://siri.org
This Web site is managed by the University of Vermont for environmental health and safety officers.

http://www.ashrae.org
This is the Web site for the American Society of Heating, Refrigeration, and Air Conditioning Engineers, Inc. It has a searchable database of information about available publications and technologies.

http://www.2nature.org
Second Nature is a nonprofit organization dedicated to promoting environmental literacy. This Web site has a searchable database of resources, including curriculum revisions to incorporate environmental issues.

# Notes

## Chapter 1

1. Although some of the dumping that caused these Superfund sites occurred prior to the creation of EPA, violations of existing laws that regulate waste handling and disposal can still be found at many schools today.

2. Hazardous wastes are generated throughout the campus from chemical use in laboratories to paints and solvents in the facilities departments and art studios. A large-quantity generator of hazardous waste is defined by the Resource Conservation and Recovery Act as an institution that generates more than 1,000 kg of hazardous wastes per month.

3. Tufts University Offices of Institutional Research and Undergraduate Admissions, *1994 New Student Survey* (Medford, MA: Tufts University, 1994).

4. See, for example, Julian Keniry, *Ecodemia* (Washington, DC: National Wildlife Federation, 1995).

## Chapter 2

1. See the models of universities proposed by Robert Birnbaum, *How Colleges Work* (San Francisco: Jossey-Bass, 1988).

2. John Nason, *Trustee Responsibilities* (Washington, DC: Association of Governing Boards of Universities and Colleges, 1989).

3. Birnbaum, *How Colleges Work,* 5.

4. "Green" Committee on Environmental Studies, *Final Report* (Muncie, IN: Ball State University, 1991), 20.

5. Patricia S. Dillon and Kurt Fischer, *Methods and Motivations for Corporate Environmental Management* (Medford, MA: Tufts University Center for Environmental Management, 1992), 36.

6. See http://iisd1.iisd.ca/educate/policybank.asp for the text of policies at other universities.

7. Presidential Executive Orders 12873 (October 20, 1993), 12845 (April 21, 1993), 12844 (April 21, 1993).

8. Jack Gault and Martha K. Potvin, "Adopting Business Strategies for Planned Social Change: A Case Study in Developing a Comprehensive 'Green' Campus Model," in *Greening of the Campus, Conference Proceedings* (Muncie, IN: Ball State University, April 1996), 185.

9. Personal Communication, July 23, 1992, Tufts Green Purchasing Forum.

10. See Julian Keniry, *Ecodemia* (Washington, DC: National Wildlife Federation, 1995), for examples of individuals who have taken leadership roles at their universities.

11. Barbara Bush, "Measuring Pollution Prevention Progress: How Do We Get There from Here?" *Pollution Prevention Review* 2,4 (Autumn 1992): 431–443.

12. Ibid.

13. Gault and Patvin, "Adopting Business Strategies," 185.

## Chapter 3

1. http://earthsystems.org/curc/curc.html provides information about the College and University Recycling Coalition of the National Recycling Coalition.

2. Sarah Hammond Creighton, *Source Reduction in Colleges and Universities: A Status Report* (Medford, MA: Tufts University Center for Environmental Management, 1993), 20.

3. http://camel.conncoll.edu/ccrec/greennet/recycling.html is the Internet address.

4. http://ecosys.drdr.virginia.edu is the Internet address.

5. http://www.wisc.edu/campus_ecology/recycle/statisti/statisti.htm is the Internet address.

6. Dartmouth College Ad Hoc Committee on Environmental Policies for College Operations, *Report of the Ad Hoc Committee on Environmental Policies for College Operations* (Hanover NH, November 1994).

7. For a discussion of this topic, see Daniel Einstein, "Comparing Campus Recyling Programs: Apples to Apples or Whales to Lawnmowers" (paper presented at the Association of Higher Education Facilities Officers Educational Conference, Salt Lake City, Utah, July 23, 1996). See http://www.wisc.edu/campus_ecology/profile/apples.htm for the full text.

8. Creighton, *Source Reduction,* 6.

9. Dartmouth College Ad Hoc Committee, *Report.*

10. http://www.ithaca.edu/orgs/env/env1/compost.html provides information about Ithaca College's composting facility.

11. U.S. Congress, Office of Technology Assessment (OTA), *Building Energy Efficiency,* OTA-E-518 (Washington, DC: U.S. Government Printing Office, May 1992), 48.

12. In order to make accurate comparisons of utility use from one year to the next, the data need to be evaluated on a consistent basis that accounts for the influences in the way the building was used. Thus, utility data might be evaluated on a per day or per student basis in order to reflect the additional demands of opening a facility more often or housing more students would put on utility use for that period.

13. OTA, *Building Energy Efficiency,* 39.

14. Decision Focus, *TAG Technical Assessment Guide,* EPRI P-4463-SR (Palo Alto, CA: Electric Power Research Institute, October 1988), vol. 2, pt. 2, pp. 5–106, in OTA, *Building Energy Efficiency,* 49.

15. U.S. EPA and U.S. Department of Health and Human Services, *Building Air Quality: A Guide for Building Owners and Facility Managers,* EPA 400/1-91/033 (Washington, DC: U.S. Government Printing Office, December 1991), 7.

16. OTA, *Building Energy Efficiency,* 37.

17. Ibid, 50.

18. Helen Kessler, Jack S. Wolpert, and Craig R. Schuttenberg, "Saving Energy Through Design Assistance at the University of Chicago Graduate School of Business," (paper presented at the Greening of the Campus Conference, Ball State University, Muncie, IN, April 4–6, 1996).

19. OTA, *Building Energy Efficiency,* 50.

20. Ibid.

21. Donna Canalos and Jim Mroz, "EPA's Green Lights Program Saves Money and Prevents Pollution," *Facilities Manager* (Spring 1995): 31.

22. Ibid.

23. T-12 fluorescent lamps are common in two-, four-, and eight-foot fixtures and can be replaced with T-8 lamps that are more efficient. Ballasts will need to be retrofitted as well, but the new ballasts usually have improved efficiency too.

24. EPA and DHHS, *Building Air Quality,* 7.

25. Ibid.

26. U.S. Environmental Protection Agency et al., *Indoor Air Quality Tools for Schools, Action Kit,* EPA 402-K-95-001 Washington, DC: U.S. Government Printing Office, September 1995.

27. See http://www.ashrae.org.

28. EPA et al., *Indoor Air Quality Tools,* 1995.

29. Diana Lopez Barrett and William D. Browning, *A Primer on Sustainable Building* (Snowmass, CO: Rocky Mountain Institute, 1995), 84.

30. OTA, *Building Energy Efficiency,* 57.

31. Weston D. Birdstall, "A Project for Improving Energy Efficiency in Residences at the University of Northern Iowa" (paper presented at Greening the Campus Conference, Ball State University, Muncie, IN, April 1996).

32. Pequod Associates, *Non-Domestic Water Audit Report, Bio-chem Laboratory,* prepared for the Massachusetts Water Resources Authority (Charlestown, MA, 1988.

33. Primary sewage treatment is a mechanical process to filter debris. Solids are settled in a sedimentation tank. Secondary treatment uses biological processes to remove biodegradable organic wastes. Tertiary or advanced treatment uses chemical and physical processes to clean the water further.

34. Julie L. Nicklin, "Deferred Maintenance: A University Tries to Repair Its Aging Buildings," *Chronicle of Higher Education,* November 8, 1996, A35–A36.

35. Oregon Energy Efficiency in Schools Task Force, *Improving Energy Efficiency in Schools* (1990), 7.

36. Green Seal is an independent, nonprofit organization dedicated to protecting the environment by promoting the manufacture and sale of environmentally responsible consumer products. It sets environmental standards and awards a Green Seal of Approval to products that cause less harm to the environment than other similar products. Address: Green Seal, 1730 Rhode Island Avenue NW, Suite 1050, Washington DC 20036 Telephone: (202)331-7337; www.greenseal.org is the Web site.

37. Joel Elkington, Julia Hailes, and Joel Makower, *The Green Consumer* (New York: Penguin Books, 1990), 121.

38. A Materials Safety Data Sheet lists only carcinogens (in any quantity) and noncarcinogens that are greater than 1 percent of the total mixture.

39. "Clean and Green," *Green Business Letter* (Washington, DC: Tilden Press, December 1993) 7.

40. Stephen Ashkin, "Housekeeping and Custodial Practices," in David Gottfried, ed., *Sustainable Building Technical Manual: Green Building Practices for Design, Construction, and Operations,* Washington, DC: U.S. Green Building Council, 1996). Order from U.S. Green Building Council, (415)543-3001, or see its web site: http://www.usgbc.org.

41. GreenLink provides an Internet bulletin board with information on used oil, motor vehicle air-conditioning, antifreeze, paints and thinners, and underground storage tanks. For further information call (888)GRN-LINK. http://www.ccar-greenlink.org is the Web page.

42. U.S. Environmental Protection Agency, *Managing Asbestos in Place,* 20T-2003. (Washington, DC: U.S. Government Printing Office, 1990), 3.

43. The EPA Asbestos hot line can provide additional information: (800)368-5888.

44. See, for example, U.S. Environmental Protection Agency, *MUSTS for USTS: A Summary of Federal Regulations for Underground Storage Tank Systems,* EPA 510-K-95-002 (Washington, DC: U.S. Government Printing Office, 1995).

45. David Saphire, *Making Less Garbage on Campus: A Hands-On Guide* (New York: Inform, 1995), 60.

46. Sheldon Cohen and Alan Pickaver, eds., *Climbing Out of the Ozone Hole* (Greenpeace International, October 1992), 12.

47. James Piper, "Replacement Chillers, Innovative Technology Equals End-User Benefits," *Building Operating Management* (January 1995).

48. Ibid.

49. See ASHRAE Guidelines 3-1990, Reducing Emissions of Fully Halogenated Chlorofluorocarbon Refrigerants in Refrigeration and Air Conditioning Equipment Applications.

50. Cohen and Pickaver, *Climbing Out,* 37.

51. Nancy Sokol Green, "America's Toxic Schools," *E, the Environmental Magazine* (November–December 1992): 33.

52. U.S. Environmental Protection Agency, *Pest Control in the School Environment: Adopting Integrated pest Management,* EPA 735-F-93-012 (Washington, DC: Government Printing Office, August 1993), 15.

53. http://camel.conncoll.edu/ccrec/greennet/campusops.html#grounds is the Internet address.

54. F. Herbert Bormann, Diana Balmori, and Gordon T. Geballe, *Redesigning the American Lawn: A Search for Environmental Harmony* (New Haven, CT: Yale University Press, 1993), 62.

55. G. Tyler Miller, Jr., *Living in the Environment* (Belmont, CA: Wadsworth, 1990), 220.

56. Robert Watson, *Case Study of Energy Efficient Building Retrofit: 40 West 20th Street, Headquarters of the Natural Resources Defense Council* (New York: NRDC, March 1990), 4.

57. See http://www.oberlin.edu/~envs/criteria.html.

58. Donald Watson, ed., *The Energy Design Handbook* (Washington, DC: American Institute of Architects, 1993), x.

59. Joseph J. Romm and William D. Browning, *Greening the Building and the Bottom Line: Increasing Productivity Through Energy-Efficient Design* (Snowmass, CO: Rocky Mountain Institute), 1994.

60. Kessler, Wolpert, and Schuttenberg, "Saving Energy."

61. Organizations such as the Architects for Social Responsibility in Boston, the American Institute of Architects (202/626-7300), and the U.S. Green Buildings Council can assist in finding professionals with the correct experience and knowledge.

62. Burk Miller Thayer, "A Passive Solar University Center," *Solar Today,* March 7, 1996, 34–37.

63. R. Watson, *Case Study,* 11.

64. D. Watson, *Energy Design Handbook,* 93.

65. Julian Keniry, *Ecodemia* (Washington, DC: National Wildlife Federation, 1995), 144.

66. For a complete listing of products and suppliers, see Andrew St. John, ed. *The Sourcebook for Sustainable Design* (Boston: Architects for Social Responsibility, 1992).

67. P. Rittelmann and P. Scanlon, "HVAC Design Delivers Twin Benefits," *Building Design and Construction,* November, 1984.

68. For a discussion of these technologies, see Leda Gawdiak Huta, ed., "Refrigeration and Air Conditioning," *Energy Ideas* 3,1 (Spring 1994): 2–12.

69. D. Watson, *Energy Design Handbook,* provides a comprehensive and detailed description of many building design opportunities.

70. *Environmental Building News,* R.R. 1, Box 161, Brattleboro, Vermont 05301, can provide an annotated bibliography of sources. Send $12 for a copy. Their Web site, http://www.ebuild.com, also has product information.

71. Thayer, "Passive Solar University Center."

72. Miller, *Living in the Environment,* 434.

## Chapter 4

1. The Massachusetts position is paid for by the Department of Environmental Protection under the Clean State Initiative Program.

2. Green Seal is an independent, nonprofit organization dedicated to protecting the environment by promoting the manufacture and sale of environmentally responsible consumer products. It sets environmental standards and awards a Green Seal of Approval to products that cause less harm to the environment than other similar products. Address: Green Seal, 1730 Rhode Island Avenue, NW, Suite 1050, Washington DC 20036 Phone: (202)331-7337. http://www.greenseal.org is the Web site.

The following Web sites are sources of information. www.metrokc.gov/oppis/recyclea.html provides information on King County, Washington, recycled products. www.epa.gov/epaoswer/non-hw/procure/comp.htm contains EPA's comprehensive procurement guidelines for recycled products. www.magnet.state.ma.us/osd/enviro/enviro.htm is the Massachusetts Environmentally Preferable Product web site.

3. Julian Keniry, *Ecodemia* (Washington, DC: National Wildlife Federation, 1995), 6.

4. Tellus Institute, *CSG/Tellus Packaging Study, Report #5, Executive Summary* (Boston: Tellus Institute for Resource and Environmental Studies, 1992), 12.

5. See http://www.wisc.edu/campus_ecology/cerp/swap/contents.htm for more information.

6. See www.epa.gov/docs/appdstar/esoe/index.html for the Energy Star home page.

7. Ibid.

8. EPA's Energy Star Program telephone is (800)STAR-YES. www.epa.gov/docs/appdstar/esoe/index.html is the Web site.

9. American Council for an Energy Efficient Economy, *Guide to Energy Efficient Office Equipment,* EPRI TR-102545-R1, (Berkeley, CA: American Council for an Energy-Efficient Economy, 1996), 38.

10. Ibid., 47.

11. Ibid., 43.

12. The National Printer's Compliance Assistance Center, maintained by the University of Illinois Champaign-Urbana, is available to provide compliance and pollution prevention assistance to the printing industry. They can be reached at (217)333-8940; or http://www.pneac.org is the Web site. EPA's *Guides to Pollution Prevention: The Commercial Printing Industry* (EPA 625790008) can also help university print shops to reduce or eliminate hazards.

13. Claudia G. Thompson, *Recycled Papers: The Essential Guide* (Cambridge, MA: MIT Press, 1992), 118.

14. Ibid.

15. William R. Moomaw, *A Northeast Response to Global Climate Change* (Medford, MA: Tufts University Center for Environmental Management, 1991).

16. Keniry, *Ecodemia,* 78.

17. Not every university will have an entire department dedicated to environment, health, and safety; however, universities must assign the administrative and operational responsibility of waste disposal, waste-related reporting, and safety oversight to an individual or department to ensure that it is handled properly. In some universities that do not have a dedicated staff, a member of the purchasing department handles this responsibility.

18. See David Saphire, *Making Less Garbage on Campus: A Hands-On Guide* (New York: Inform, 1995), 60–63, for examples of environmentally friendly paints and adhesives.

19. "Household Batteries: Management Dilemma," *Household Hazardous Waste Management News* (Summer 1989): 3.

20. American Chemical Society, Task Force on RCRA, *Less Is Better* (Washington DC: American Chemical Society, 1985).

## Chapter 5

1. Mikael Backman, "Energy in the Food-Producing System," in A. V. Bridgwater and K. Lidgren, eds., *Energy in Packaging and Waste* (London: Van Nostrand Reinhold Co., 1983), 33.

2. Society for Foodservice Management, "Polystyrene vs. Polycarbonate: A Matter of Fact or Fiction?" in *The Annual* (1991–1992): 83.

3. "UCLA Will Switch to Polystyrene Servingware," *Food Service Director* (November 15, 1991), 1.

4. McDonald's Corporation and Environmental Defense Fund Waste Reduction Task Force, *Final Report* (1991), 39.

5. Tellus Institute, *CSG/Tellus Packaging Study, Report #5, Executive Summary* (Boston: Tellus Institute for Resource and Environmental Studies, 1992), 12.

6. Ibid, 24, 43.

7. Caroline Ganley and Peter Allison, "Analysis of Dining Hall Waste Sort," in Sarah Hammond Creighton, *Source Reduction in Colleges and Universities* (Medford, MA: Tufts University Center for Environmental Management, 1993), 53.

8. U.S. Congress Office of Technology Assessment. *Facing America's Trash, What Next for Municipal Solid Waste?* OTA-O-424. Washington, D.C.: U.S. Government Printing Office. (October, 1989): 80.

9. McDonald's Corporation and Environmental Defense Fund Waste Reduction Task Force, *Final Report,* 31.

10. Ganley and Allison, "Analysis," 59.

11. Ibid.

12. Ann M. Messersmith, George M. Wheeler, and Victoria Rousso, "Energy Used to Produce Meals in School Food Service," *School Food Service Research Review* 18, 1 (1994): 35.

13. See Ann M. Messersmith,"School Food Service Energy," *School Food Service Research Review,* 18, 1 (1994): 39–44, for a bibliography of sources about energy savings in dining services.

14. Mardee Haidin Regan, "Kitchen Spy: Electrotech on Trial," *Food Arts,* (July–August 1995): 86–89.

15. For comparative purposes, 1 kWh (measure of electricity use) equals 3,413 BTUs (measure of gas use).

16. Regan, "Kitchen Spy," 1995.

17. Architectural Energy Corporation and Pennsylvania State University, *The Model Electric Restaurant,* Vol. 2: *Supporting Data for Energy and Economic Analysis,* prepared for the Electric Power Research Institute, EPRI CU-6702 (December 1990), G-3.

18. Based on Sarah Hammond Creighton, "Environmental Action in University Food Services," in *Proceedings of the Greening of Universities Conference,* Ball State University, April 1996, and Eric Friedman and Sarah Hammond Creighton, "Tufts Dining Energy Efficiency Model Program, Final Report," unpublished ms. (Medford, MA: Tufts University, 1994).

19. Architectural Energy Corporation and Pennsylvania State University, *Model Electric Restaurant,* B24-B39.

20. EPA's Safe Drinking Water Hotline can provide important information and answer questions about water quality: (800)426-4791.

21. Gary L. Valen, "Hendrix College Local Food Project," in David Orr and David Eagan, eds., *Campus and Environmental Responsibility,* New Directions for Higher Education, no. 77 (San Francisco: Jossey-Bass, Spring 1992), 86.

22. Representative from Shaheen Brothers, Inc. at Tufts Dining Conference, September 1993, Medford MA.

23. Meat and dairy are two to three times as energy intensive as the production of vegetables according to Backman, "Energy in the Food-Producing System," 33.

24. EarthSave, *Our Food Our World* (Santa Cruz, CA: EarthSave Foundation, 1992).

## Chapter 6

1. Franklin Associates, *Characterization of Municipal Solid Waste in the United States: 1990 Update* (Washington, DC: U.S. Environmental Protection Agency Office of Solid Waste, June 1990), 81.

2. Sarah Hammond Creighton, *Source Reduction in Colleges and Universities: A Status Report* (Medford, MA: Tufts University Center for Environmental Management, 1993), 6.

3. Daniel Einstein communication to Grnsch-l@brownvm.brown.edu user group, November 17, 1995.

4. William Rathje and Cullen Murphy, *Rubbish! The Archaeology of Garbage* (New York: HarperCollins, 1992), 243.

5. Julian Keniry, *Ecodemia* (Washington, DC: National Wildlife Federation, 1995), 125.

6. Louis A. Carrière and Mark S. Rea, "Economics of Switching Fluorescent Lamps," *IEE Transactions on Industry Applications* 24, 3 (May–June 1988), 370.

7. To receive information about sending books overseas, contact the International Book Project in Lexington, Kentucky, at (606)254-6771.

8. This section was based in part on Lucy Edmondson and Sarah Hammond Creighton, "Transportation and Parking at Tufts University: Opportunities to Meet Obligations and Improve the Environment," unpublished ms. (Medford, MA: Tufts University, Center for Environmental Management, 1992).

9. Cornell University Office of Transportation Services, *Commuting Solutions, Summary of Transportation Demand Management Program (TDMP),* rev. (Ithaca, NY, June 1996).

10. Keniry, *Ecodemia,* 47.

11. E. Scott Geller, "Applied Behavior Analysis and Social Marketing: An Integration for Environmental Preservation," *Journal of Social Issues* 45 (November 1989): 20; Frederick Buttell, "New Directions in Environmental Sociology," *Annual Review of Sociology* 13 (1987): 475.

12. Cornell University Office of Transportation Services, *Community Solutions.*

13. David Orr, *Ecological Literacy: Education and the Transition to a Postmodern World* (Albany, NY: State University of New York Press, 1992), 90.

14. Anthony Cortese, "Education for an Environmentally Sustainable Future," *Environment Science and Technology* 26: 6 (1992): 1108–1114.

15. Anthony Cortese now heads Second Nature, a non-profit organization dedicated to promoting environmental literacy. See http://www.2nature.org for more information.

16. Orr, *Ecological Literacy,* 133.

17. Ibid.

18. See http://www.wisc.edu/campus_ecology/cerp.htm for a list of research projects and related initiatives.

## Chapter 7

1. P. C. Ashbrook and P. A. Reinhardt, "Hazardous Waste in Academia." *Environmental Science and Technology* 19(1985):1150–1155.

2. For an overview of environmental regulations that apply to colleges and universities, see National Association of College and University Business Officers (NACUBO), *Hazardous Waste Management at Educational Institutions* Washington, DC: NACUBO, 1987), and Paul G. Wallach et al., *Environmental Requirements for Colleges and Universities* (Washington, DC: Hale and Dorr, 1991).

3. See NACUBO, *Hazardous Waste Management,* 79–82, for a sample training program.

4. National Microscale Chemistry Center, Merrimack College, 315 Turnpike Street, North Andover, MA 01845, (508)837-5137.

5. Ralph Stone and Co., *Waste Audit Study, Research and Educational Institutions* (Los Angeles: California Department of Health Services, 1988).

6. Jamie Spigler, "Waste Not Want Not," *American School and University* (February 1992): 64a.

7. Joint Council for Health, Safety, and Environmental Education of Professionals, *Guidelines for Incorporating Safety and Health into Engineering Curricula,* Vol. 1: *Laboratory Safety* (Savory, IL: Joint Council, 1994).

8. Ventilation from fume hoods may run counter to energy-efficiency efforts. There are technologies that can help achieve efficient ventilation and exhaust, such as heat exchangers and shut-off systems.

9. For a complete list of rules, see Joint Council, *Guidelines,* 1: 11–12.

10. Very small quantity generators of hazardous waste are defined as those that generate less than 100 kg per month.

11. Flinn Scientific, *Flinn Chemical Catalog Reference Manual* (Batavia, IL, 1994).

12. John A. Valicenti, "Keeping the Lid On," *American School and University* (April 1992): 32o–32p.

13. Tufts University, *Hazardous Waste Management in Educational Institutions: A Report to the United States Environmental Protection Agency* (Medford, MA: Center for Environmental Management, Tufts University, 1987).

14. Ibid.

15. In cooperation with the American Chemical Society, some colleges and universities are proposing federal regulations to allow on-site pretreatment of some wastes. Even if this practice is allowed, it should be undertaken with caution.

16. Julian Keniry, *Ecodemia* (Washington, DC: National Wildlife Federation, 1995), 171.

17. Steven Pike, "Little-Known Hazards to Watch For," *American School and University*, (June 1990), 21, 24.

18. A large-quantity generator of hazardous wastes generates more than 1,000 kg of waste per month.

## Chapter 8

1. See "Blueprint for a Green Campus: The Campus Earth Summit for Higher Education," January 1995. Yale University. http://www.nwf.org/campus.

2. Student Environmental Action Coalition (SEAC) P.O. Box 248 Tucson, AZ 85702, (520)903-0128, provides assistance and organizing to students interested in the environmental movement. http://www.seac.org is the Web site.

3. E. Scott Geller, "Applied Behavior Analysis and Social Marketing: An Integration for Environmental Preservation," *Journal of Social Issues* 45 (November 1989): 20; Frederick Buttell, "New Directions in Environmental Sociology," *Annual Review of Sociology* 13 (1987): 475.

4. See http://www.nwf.org/campus.

## Chapter 9

1. Following a staff change, Tufts has now implemented a water-conserving irrigation system on the athletic fields.

2. David Orr, "The Problem of Education" in David Eagan and David Orr, eds., *The Campus and Environmental Responsibility* (San Francisco: Jossey-Bass, 1992), 4–5.

3. Wendell Berry, "Out of Your Car, Off Your Horse," *Land Stewardship Letter* (Marine, MN) (Summer 1991).

# Bibliography

American Chemical Society, Task Force on RCRA. *Less Is Better*. Washington, DC: American Chemical Society, 1985.

American Council for an Energy-Efficient Economy. *Guide to Energy Efficient Office Equipment*. EPRI TR-102545-R1. Berkeley, CA: American Council for an Energy-Efficient Economy, 1996.

Architectural Energy Corporation and the Pennsylvania State University. *The Model Electric Restaurant*. Vol. 2: *Supporting Data for Energy and Economic Analysis*. Prepared for the Electric Power Research Institute, EPRI CU-6702, December 1990.

Ashbrook, P. C., and P. A. Reinhardt. "Hazardous Waste in Academia." *Environmental Science and Technology* 19 (1895): 1150–1155.

Ashkin, Stephen. "Housekeeping and Custodial Practices," In David Gottfried, ed., *Sustainable Building Technical Manual: Green Building Design, Construction, and Operations*. San Francisco: Public Technology, 1996.

Backman, Mikael. "Energy in the Food-Producing System." In A. V. Bridgwater and K. Lidgren, eds., *Energy in Packaging and Waste*. London: Van Nostrand Reinhold, 1983.

Barrett, Diana Lopez, and William D. Browning. *A Primer on Sustainable Building*. Snowmass, CO: Rocky Mountain Institute, 1995.

Berry, Wendell. "Out of Your Car, Off Your Horse." *Land Stewardship Letter* (Marine, MN) (Summer 1991).

Birnbaum, Robert. *How Colleges Work*. San Francisco: Jossey-Bass, 1988.

Bormann, F. Herbert, Diana Balmori, and Gordon T. Geballe. *Redesigning the American Lawn: A Search for Environmental Harmony*. New Haven, CT: Yale University Press, 1993.

Buntell Frederick. "New Directions in Environmental Sociology." *Annual Review of Sociology* 13 (1987): 465–488.

Bush, Barbara. "Measuring Pollution Prevention Progress: How Do We Get There from Here?" *Pollution Prevention Review* 2, 4 (Autumn 1992): 431-443.

"Blueprint for a Green Campus: The Campus Earth Summit for Higher Education," January, 1995. Yale University. http://www.nwf.org/campus

Canalos, Donna, and Jim Mroz. "EPA's Green Lights Program Saves Money and Prevents Pollution." *Facilities Manager* (Spring 1995): 31.

Carrière, Louis A., and Mark S. Rea. "Economics of Switching Fluorescent Lamps." *IEEE Transactions on Industry Applications* 24, 3 (May–June 1988): 370–379.

"Clean and Green." *Green Business Letter* (December 1993).

Clean Air Act Amendments of 1990: Detailed Summary of Titles. US EPA. November 30, 1990.

Cohen, Sheldon, and Alan Pickaver, eds. *Climbing out of the Ozone Hole: A Preliminary Survey of Alternatives to Ozone-Depleting Chemicals.* Greenpeace International, October 1992.

Cornell University Office of Transportation Services. *Commuting Solutions: Summary of Transportation Demand Management Program (TDMP).* Revised. Ithaca, NY, June 1996.

Cortese, Anthony. "Education for an Environmentally Sustainable Future." *Environment Science and Technology* 26, 6 (1992): 1108–1114.

Creighton, Sarah Hammond. "Environmental Action in University Food Services." In *Proceedings of the Greening of Universities Conference.* Muncie, IN: Ball State University, April 1996.

———. *Source Reduction in Colleges and Universities: A Status Report.* Medford, MA: Tufts University Center for Environmental Management, 1993.

Dartmouth College Ad Hoc Committee on Environmental Policies for College Operations. *Report of the Ad Hoc Committee on Environmental Policies for College Operations.* Hanover, NH, November 1994.

Dartmouth College Energy Council. *Dartmouth College Energy Policy.* Hanover, NH, n.d.

———. *New Construction Energy Standards.* Hanover, NH, n.d.

Decision Focus. *TAG Technical Assessment Guide.* EPRI P-4463-SR. Palo Alto, CA: Electric Power Research Institute, October 1988.

Dillon, Patricia, and Kurt Fischer. *Methods and Motivations for Corporate Environmental Management.* Medford, MA: Tufts University Center for Environmental Management, 1992.

Eagan, David E., and David W. Orr, eds. *The Campus and Environmental Responsibility.* New Directions for Higher Education, 77. San Francisco: Jossey-Bass, 1992.

EarthSave. *Our Food Our World.* Santa Cruz, CA: EarthSave Foundation, 1992.

Edmondson, Lucy, and Sarah Hammond Creighton. "Transportation and Parking at Tufts University: Opportunities to Meet Obligations and Improve the Environment." Unpublished ms. Medford, MA: Tufts University, Center for Environmental Management, 1992.

Einstein, Daniel. "Comparing Campus Recyling Programs: Apples to Apples or Whales to Lawnmowers." Paper presented at the Association of Higher Education Facilities Officers Educational Conference, Salt Lake City, Utah, July 23, 1996.

Elkington, Joel, Julia Hailes, and Joel Makower. *The Green Consumer*. New York: Penguin Books, 1990.

Flinn Scientific. 1994. *Flinn Chemical Catalog Reference Manual*. Batavia, IL.

Foley, Joseph E. "Management Strategies Keep Compliance in Check." *American School and University* (December 1991): pp 36b–36f.

*Food Service Director*. "UCLA Will Switch to Polystyrene Servingware." November 15, 1991.

Franklin Associates. *Characterization of Municipal Solid Waste in the United States: 1990 Update*. Washington, DC: U.S. Environmental Protection Agency, Office of Solid Waste, June 1990.

Friedman, Eric. "Measuring What Isn't: A Framework for Evaluating the Impacts of Municipal Solid Waste Source Reduction Programs." Master's thesis, Tufts University, 1994.

Friedman, Eric, and Sarah Hammond Creighton. "Tufts Dining Energy Efficiency Model Program, Final Report." Unpublished ms. Medford, MA: Tufts University, 1994.

Ganley, Caroline, and Peter Allison. "Analysis of Dining Hall Waste Sort." In Sarah Hammond Creighton, *Source Reduction in Colleges and Universities*. Medford, MA: Tufts University Center for Environmental Management, 1993.

Gault, Jack, and Martha K. Potvin. "Adopting Business Strategies for Planned Social Change: A Case Study in Developing a Comprehensive 'Green' Campus Model." In *Greening of the Campus, Conference Proceedings*. Muncie, IN: Ball State University, Muncie, IN, April 1996.

Geller, E. Scott. "Applied Behavior Analysis and Social Marketing: An Integration for Environmental Preservation." *Journal of Social Issues* 45 (November 1989): 20.

Gottfried, David, ed. *Sustainable Building Technical Manual: Green Building Design, Construction, and Operations*. San Francisco: Public Technology, 1996.

The "Green" Committee on Environmental Studies. *Final Report*. Muncie, IN: Ball State University, 1991.

Green, Nancy Sokol. "America's Toxic Schools." *E, the Environmental Magazine,* 3 (November–December 1992): 30–37.

Green Seal, Inc. *Campus Green Buying Guide*. Washington, DC: Green Seal, 1994.

"Household Batteries: Management Dilemma." *Household Hazardous Waste Management News* (Andover, MA) (Summer 1989).

Huta, Leda Gawdiak, ed. "Refrigeration and Air Conditioning." *Energy Ideas* 3, 1 (Spring 1994): 2–12.

Hynes, H. Patricia. *Earth Right.* Rocklin, CA: Primas Publishing and Communications, 1990.

Joint Council for Health, Safety, and Environmental Education of Professionals. *Guidelines for Incorporating Safety and Health into Engineering Curricula.* Vol. 1: *Laboratory Safety.* Savory, IL: Joint Council, 1994.

Kaufman, James A., ed. *Waste Disposal in Academic Institutions.* Chelsea, MI: Lewis Publishers, 1990.

Keniry, Julian. *Ecodemia: Campus Environmental Stewardship at the Turn of the 21st Century.* Washington, DC: National Wildlife Federation, 1995.

Kessler, Helen, Jack S. Wolpert, and Craig R. Schuttenberg. "Saving Energy Through Design Assistance at the University of Chicago Graduate School of Business." In *Proceedings of the Greening of the Campus.* Ball State University, Muncie, IN, April 4–6, 1996.

"Lawn Care Industry Dilemma." *American Horticulturist* 69, 11 (1990): 4.

Lewis, Eleanor J., and Eric Weltman. *Forty Ways to Make Government Purchasing Green.* Washington, DC: Center for Study of Responsive Law, 1992.

Massachusetts Bar Association. *Environmental Problems Facing Colleges and Universities in the 1990's: Meeting the Challenge.* Cambridge, MA, April 1990.

Mayo, Dana W., Ronald M. Pike, and Samuel S. Butcher. *Microscale Organic Laboratory.* New York: Wiley, 1986.

Mayo, Dana W., Ronald M. Pike, Samuel S. Butcher, and Peter K. Trumper. *Microscale Techniques for the Organic Laboratory.* New York: Wiley, 1991.

McCann, Michael. *Artist Beware.* New York: Watson- Guptill, 1979.

McDonald's Corporation and Environmental Defense Fund Waste Reduction Task Force. *Final Report.* 1991.

Messersmith, Ann M. "School Food Service Energy." *School Food Service Research Review* 18, 1 (1994): 39–44.

Messersmith, Ann M., George M. Wheeler, and Vitoria Rousso. "Energy Used to Produce Meals in School Food Service." *School Food Service Research Review* 18, 1 (1994): 29–37.

Miller, G. Tyler, Jr. *Living in the Environment.* Belmont, CA: Wadsworth, 1990.

Moomaw, William R. *A Northeast Response to Global Climate Change.* Medford, MA: Tufts University Center for Environmental Management, 1991.

Nason, John. *Trustee Responsibilities.* Washington, DC: Association of Governing Boards of Universities and Colleges, 1989.

National Association of College and University Business Officers (NACUBO). *Hazardous Waste Management at Educational Institutions.* Washington DC: NACUBO, 1987.

National Research Council. *Prudent Practices for Disposal of Chemicals from Laboratories.* New York: Van Nostrand Reinhold, 1989.

National Roundtable on the Environment and the Economy. *The Green Guide: A User's Guide to Sustainable Development for Canadian Colleges.* Ottowa, Ontario, 1992.

Nelson, John Olaf. "Water Conserving Landscapes Show Impressive Savings." *Management and Operations Journal AWWA* (March 1987): 35–42.

Nicklin, Julie L. "Deferred Maintenance: A University Tries to Repair Its Aging Buildings." *Chronicle of Higher Education,* November 8, 1996, A35–A36.

Oregon Energy Efficiency in Schools Task Force established by Oregon State University Extension Service, Oregon Department of Energy, and Oregon Department of Education. *Improving Energy Efficiency in Schools.* 1990.

Orr, David W. *Ecological Literacy: Education and the Transition to a Postmodern World.* Albany, NY: State University of New York Press, 1992.

Pequod Associates. *Non-Domestic Water Audit Report, Bio-chem Laboratory.* Prepared for the Massachusetts Water Resouces Authority. Charlestown, MA, 1988.

Perrin, Noel. "Colleges Are Doing Pitifully Little to Protect the Environment." *Chronicle of Higher Education,* October 22, 1992, B4–B5

Pike, Steven. "Little-Knowns Hazards to Watch For." *American School and University* (June 1990): 21, 24.

Piper, James. "Replacement Chillers, Innovative Technology Equals End-User Benefits." *Building Operating Management* (January 1995).

Ralph B. Stone and Co. *Waste Audit Study, Research and Educational Institutions.* Los Angeles: California Department of Health Services, 1988.

Rathje, William, and Cullen Murphy. *Rubbish! The Archaeology of Garbage.* New York: HarperCollins, 1992.

Regan, Mardee Haidin. "Kitchen Spy: Electrotech on Trial." *Food Arts* (July–August 1995): 86–89.

Rittelman, P., and P. Scanlon. "HVAC Design Delivers Twin Benefits." *Building Design and Construction* (November 1984).

Robbins, John. *Diet for a New America.* Walpole, NH: Stillpoint Publishing, 1987.

Romm, Joseph J., and William D. Browning. *Greening the Building and the Bottom Line: Increasing Productivity Through Energy-Efficient Design.* Snowmass, CO: Rocky Mountain Institute, 1994.

St. John, Andrew, ed. *The Sourcebook for Sustainable Design.* Boston: Architects for Social Responsibility, 1992.

Saphire, David. *Making Less Garbage on Campus: A Hands-On Guide.* New York: Inform, 1995.

Saunders, Tedd, and Loretta McGovern. *The Bottom Line of Green Is Black.* New York: HarperCollins, 1993.

Smith, April A., and the Student Environmental Action Coalition. *Campus Ecology: A Guide to Assessing Environmental Quality and Creating Strategies for Change.* Venice, CA: Living Planet Press, 1993.

Society for Foodservice Management. "Polystyrene vs. Polycarbonate: A Matter of Fact or Fiction?" *The Annual* (1991–1992): 83–87.

Spigler, Jamie. "Waste Not Want Not." *American School and University* (February 1992): 64a.

Steinman, David. *Diet for a Poisoned Planet.* New York: Ballantine Books, 1990.

Szafran, Zvi, Ronald M. Pike, and Mono M. Singh. *Microscale Inorganic Chemistry: A Comprehensive Laboratory Experience.* New York: Wiley, 1991.

Tellus Institute. *CSG/Tellus Packaging Study, Report #5, Executive Summary.* Boston: Tellus Institute for Resource and Environmental Studies, 1992.

Thayer, Burk Miller. "A Passive Solar University Center." *Solar Today,* March 7, 1996, 34–37.

Thompson, Claudia G. *Recycled Papers, the Essential Guide.* Cambridge, MA: MIT Press, 1992.

Tufts University. *Hazardous Waste Management in Educational Institutions: A Report to the United States Environmental Protection Agency.* Medford, MA: Center for Environmental Management, Tufts University, 1987.

Tufts University Offices of Institutional Research and Undergraduate Admissions. 1994. *New Student Survey.* Medford, MA: Tufts University, 1994.

"UCLA Will Switch to Polystyrene Servingware." *Food Service Director,* November 15, 1991, 1.

U.S. Congress. Office of Technology Assessment. *Building Energy Efficiency.* OTA-E-518. Washington, DC: U.S. Government Printing Office, May 1992.

———. *Facing America's Trash: What Next for Municipal Solid Waste?* OTA-0-424. Washington, DC: U.S. Government Printing Office, October 1989.

U.S. Environmental Protection Agency. *Business Guide for Reducing Solid Waste.* EPA/530-K-92-004. Washington, DC: U.S. Government Printing Office, 1989.

———. *Decision-Makers Guide to Solid Waste Management.* EPA/530-SW-89-072. Washington, DC: U.S. Government Printing Office, 1989.

———. *Guides to Pollution Prevention, Research and Educational Institutions.* EPA 625/7-90/010. Washington, DC: U.S. Government Printing Office, 1990.

———. *Guides to Pollution Prevention: The Commercial Printing Industry.* EPA 625/7-90/008. Washington, DC: U.S. Government Printing Office, 1990.

———. *The Inside Story: A Guide to Indoor Air Quality.* EPA 402-K-93-007. Washington, DC: U.S. Government Printing Office, 1993.

———. *Managing Asbestos in Place.* 20T-2003. Washington, DC: U.S. Government Printing Office, 1990.

———. *MUSTS for USTS: A Summary of Federal Regulations for Underground Storage Tank Systems.* EPA 510-K-95-002. Washington, DC: U.S. Government Printing Office, 1995.

———. *Pest Control in the School Environment: Adopting Integrated Pest Management*. EPA 735-F-93-012. Washington, DC: U.S. Government Printing Office, August 1993.

———. *Procurement Guidelines for Government Agencies*. EPA/530-SW-91-011. Washington, DC: U.S. Government Printing Office, December 1990.

———. *Radon Measurement in Schools*. EPA/402-R-92-014. Washington, DC: U.S. Government Printing Office, 1993.

———. *Radon Prevention in the Design and Construction of Schools and Other Large Buildings*. EPA/635-R-92-016. Washington, DC: U.S. Government Printing Office, 1993.

———. *Report to Congress: Management of Hazardous Wastes from Educational Institutions*. EPA/530-SW-89-040. Washington, DC: U.S. Government Printing Office, April 1989.

———. *Running a Conference as a Clean Product*. EPA/600/2-91-026. Washington, DC: U.S. Government Printing Office, 1991.

U.S. Environmental Protection Agency, American Federation of Teachers, Association of School Business Officials, Council for American Private Education, National Education Association, National Parent Teacher Association, and the American Lung Association. *Indoor Air Quality Tools for Schools, Action Kit*. EPA 402-K-95-001. Washington, DC: U.S. Government Printing Office, September 1995.

U.S. Environmental Protection Agency and U.S. Department of Health and Human Services. *Building Air Quality: A Guide for Building Owners and Facility Managers*. EPA 400/1-91/033. Washington, DC: U.S. Government Printing Office, December 1991.

Valen, Gary L. "Hendrix College Local Food Project." In David Eagan and David Orr eds., *The Campus and Environmental Responsibility*. San Francisco: Jossey-Bass, 1992.

Valicenti, John A. "Keeping the Lid On." *American School and University* (April 1992): 32o–32p.

Wallach, Paul G., Mark Atlas, Kenneth R. Meade, Joseph C. Stanko, Jr., and Mary B. Griffin. *Environmental Requirements for Colleges and Universities*. Washington, DC: Hale and Dorr, January 1991.

Washington State Department of Ecology. *Purchasing for the Environment: Guidelines for Developing Procurement Programs and Policies* 92-15. Olympia, WA.

Watson, Donald, ed. *The Energy Design Handbook*. Washington, DC: American Institute of Architects, 1993.

Watson, Robert. *Case Study of Energy Efficient Building Retrofit: 40 West 20th Street, Headquarters of the Natural Resources Defense Council*. New York: Natural Resources Defense Council, March 1990.

# Index

Reinsch Library
Marymount University
2807 N Glebe Road
Arlington, Virginia 22207